Wastewater: Techniques for Evaluation and Management

Wastewater: Techniques for Evaluation and Management

Editor: Gabriel Craig

MURPHY & MOORE

www.murphy-moorepublishing.com

Murphy & Moore Publishing,
1 Rockefeller Plaza,
New York City, NY 10020, USA

Visit us on the World Wide Web at:
www.murphy-moorepublishing.com

ISBN: 978-1-63987-567-2 (Hardback)

Cataloging-in-Publication Data

Wastewater : techniques for evaluation and management /
edited by Gabriel Craig.
p. cm.
Includes bibliographical references and index.
ISBN 978-1-63987-567-2
1. Sewage. 2. Sewage--Analysis. 3. Sewage--Purification. I. Craig, Gabriel.
TD735 .W378 2022
628.161--dc23

Table of Contents

Preface

It is often said that books are a boon to mankind. They document every progress and pass on the knowledge from one generation to the other. They play a crucial role in our lives. Thus I was both excited and nervous while editing this book. I was pleased by the thought of being able to make a mark but I was also nervous to do it right because the future of students depends upon it. Hence, I took a few months to research further into the discipline, revise my knowledge and also explore some more aspects. Post this process, I begun with the editing of this book.

Water which has been used for various domestic, industrial and agricultural purposes, and is not suitable for human consumption is termed as wastewater. Metals, plastics and human excreta are some of the common water pollutants. The processes used to remove these contaminants from water so that it can be returned to the water cycle with minimum environmental impact are studied under the discipline of wastewater treatment. Water treatment techniques include grit removal, settling, aeration, filtration, disinfection, and sludge treatment. Some of the different types of wastewater treatment plants are sewage treatment plants, Leachate treatment plants and agricultural wastewater treatment plants. This book attempts to understand the multiple branches that fall under the discipline of wastewater evaluation and management and how such concepts have practical applications. It elucidates new techniques and their applications in a multidisciplinary manner. As this field is emerging at a rapid pace, the contents of this book will help the readers understand the modern concepts and applications of the subject.

I thank my publisher with all my heart for considering me worthy of this unparalleled opportunity and for showing unwavering faith in my skills. I would also like to thank the editorial team who worked closely with me at every step and contributed immensely towards the successful completion of this book. Last but not the least, I wish to thank my friends and colleagues for their support.

Editor

Parasitological Contamination in Organic Composts Produced with Sewage Sludge

Eduardo Robson Duarte[1*], Flávia Oliveira Abrão[1,]
Neide Judith Faria de Oliveira[1] and Bruna Lima Cabral[1]
[1]Instituto de Ciências Agrárias da Universidade Federal de Minas Gerais
Montes Claros,
Brazil,

1. Introduction

The activities performed daily by humans generate large volumes of waste from various areas, increasing environmental pollution and public health problems. According to the recommendations of the World Health Organization (WHO), the reactor effluent treatment from domestic sewages can be used in agriculture, since it applied to cultures that provide little risk of contamination with pathogens (Ayres & Mara, 1996). The sewage sludge is classified as class A when origins of processes with effective reduction of pathogens and can be used without restrictions in horticulture, and in the class B if results of processes of moderate reduction of pathogens, with more restricted use and be applied in reforestation and other cultures in which the risk of environmental and human contamination can be better controlled (Fernandes, 2000; David, 2002).

Sewage sludge must have characteristics that allow its setting within the parameters set for each class (David, 2002). For class A, the most probable number of coliforms per gram of dried sludge must be less than 1000, and the parasitic contamination should be less than one viable egg of helminths in four grams of dried sludge and less than one egg per litre of effluent (WHO, 1989; Fernandes, 2000).

The processes most often employed for the stabilization of sewage include the aerobic and anaerobic digestion. The application of lime and the composting are also recommended in some countries like USA, France and Brazil. However, the efficiency of the stabilization processes depends of the operational quality and of the pathogen characteristics present in the sewage sludge (Paulino et al., 2001).

2. Parasites presented in sewage sludge

Several countries have researched alternatives for final disposal of the waste from water sewage and sludge treatment. The sewage, prior to the stabilization treatment and disinfection, can contain macro and micronutrients and many pathogenic microorganisms and parasites. The handling and use of sewage and sludge obtained, without prior treatment, may promote severe infection to humans and animals (Paulino et al., 2001). According to WHO, 25% of the world's hospital beds are occupied by patients with diseases transmitted by contaminated water. About 1.5 billion people are infected with *Ascaris lumbricoides* and 1.3 billion with *Ancylostoma* spp. infection (Crompton, 1999).

The establishment of maximum concentrations of viable eggs of helminths in sewage sludge has been the criterion worldwide used to allow the agricultural use of this material (Capizzi-Banas & Schwartzbrod, 2001). Epidemiological studies have shown that the high frequency of helminths in human population, the long survival of the eggs in the environment and the low infectious dose are risks associated with agricultural use of sewage sludge (Soccol et al., 1997). In the world 3.5 billion people are infected with helminths or protozoa and children are the most frequently contaminated, resulting in approximately 60 000 deaths associated with *Ascaris lumbricoides* each year and 70 000 with *Entamoeba histolytica* (WHO, 2000). Table 1 showed the main pathogenic helminths eliminated in human faeces.

Helminths	diseases	cycle characteristics
Ancylostoma duodenale	ancylostomiasis	human – soil – human
Ascaris lumbricoides	ascaridiasis	human – soil – human
Necator americanus	necatorosis	human – soil – human
Trichuris trichiura	tricurosis	human – soil – human
Taenia saginata	taeniasis	human – cattle – human
Taenia solium	taeniasis	human – pigs – human
Taenia solium	cysticercosis	human – soil - human

Roque, (1997).

Table 1. Main pathogenic helminths from human faeces

3. Sewage sludge stabilization

The sludge stabilization is used primarily to reduce substantially the numbers of pathogenic organisms, to minimize health risks, controlling odours and to restrict the possibility of further decomposition. The stability is generally associated with the tendency to biodegrade organic matter. This step is an important in sewage treatment and influences significantly the characteristics of sludge produced (Fernandes, 2001).

The resulting sludge from biological treatment systems of wastewater is composed of live microorganisms. As the efficiency of biological processes is related to the amount of living cells, active in the process, the processing systems keeps the affluent rich in a medium with sludge. A biological process is considered efficient and economical when can be operated with low hydraulic retention times and long retention times of solids to allow the growth of microorganisms. The sludge is thus an initiating factor for the processes of biological treatment of sewage and its excess is regarded as a waste. The efficiency of stabilization and production of this waste depend of the technology of sewage treatment systems and its operationalization (Fernandes 2001).

Stabilisation processes can be divided on biologic and chemist process. Biological methods involve digestion anaerobic, aerobic digestion, autothermal aerobic digestion and composting. The alkaline is the most used chemical process and usually the lime is mixed with the material, improving pH that inactivates some pathogenic microorganisms. Other chemical agents can be chlorine ozone, hydrogen peroxide and potassium permanganate but to a small scale treatment.

4. Composting and parasitic contamination reduction

According to Metcalf & Eddy (1991), composting is an economically and environmentally correct alternative to the stabilization of wastes from agriculture and industries production and urban sewage treatment systems, with possible agronomic use of such waste. This technology involves the organic waste processing into humified composts through two distinct phases. The first corresponds to biochemical reactions of oxidation and the second to the process of humification or maturation phase. Being developed by a microbial ecosystem, the composting is directly influenced by factors that act on microbial activity. Among these, the most important are aeration, temperature, humidity content and nutrient concentration (Veras & Povinelli, 2004).

Large number of treatment plants in Brazil employs the natural composting process, in which organic material, after separation of insensitive material for composting, is arranged in stacks in the courts and are periodically raked to improve aeration. However, there are some cases using expedited procedures with injection of air into the piles (Barrier et al., 2006).

5. Parasitological contamination in compounds with sewage sludge before and after thermal treatment

The temperature is an important factor to contamination reduction in the composting. Study conducted in Brazil had as objective to evaluate the contamination by eggs, cysts and oocysts of protozoa in organic compounds, using sewage sludge and agricultural residues, before and after heat treatment (Duarte et al., 2008). In the experiment conducted in Montes Claros, Minas Gerais, Brazil, 25 different organic compounds were prepared, with approximate volume of 0.5 cubic meters. The materials were kept during the first fifteen days, and daily irrigated and raked (Duarte et al., 2008).

The compounds were prepared with carbon /nitrogen relation of 25:1, 30:1, 35:1, 40:1 and 45:1. In the composition were used cotton waste (5.8 to 10%), seeds of maize and beans (11.8 to 17.2%), and grass seeds (4.4 to 4.7 %). Were also used coconut fibber (50-20%), rice husk (7.4 to 17%), chopped elephant grass (*Pennisetum purpureum*) (1.93 to 17.6%), shell of *Cariocar brasiliense* (pequi) (0.6 to 2.27%), fresh cattle manure (2.3 to 4.8%) and sewage sludge (28.8 to 38.1%). Faeces were from 40 Holstein cows with an average of 5.5 years. Each compound received 220g of rock phosphate, 21kg of gypsum and 21kg of limestone (Duarte et al., 2008). First two samples of 500g sewage sludge were collected from the Juramento Town, in North of Minas Gerais, Brazil. The sludge was collected in the dry bed of the sewage treatment station of this local that uses flow anaerobic reactor ascending. The physical and chemical analysis of sewage sludge were performed according to techniques recommended by Tedesco et al. (1995) and showed pH-H_2O 4.8 and levels of nitrogen, carbon and humidity of 2.4, 9.4 and 6.0%, respectively (Duarte et al., 2008).

Subsequently, were collected 500g samples from compounds with fifteenth day of composting, avoiding contamination from one compound to another and put them in plastic bags, clean and free from contamination by parasites. After this period, the compounds were transferred to a greenhouse with a controlled temperature of 60°C during 12 hours and subsequently to keep 55°C until the twenty-second day after treatment. This procedure was employed with the purpose of promoting the reduction of pathogens parasites of the compounds. The effectiveness of heat treatment and their uniformity was checked in all the materials produced and three different samples from several parts of these compounds corresponding to the parts A, B and C, were collected after twenty-two days of thermal treatment. Was obtained 500g for each sample (A, B and C), using the same hygienic measures to prevent the contamination (Duarte et al., 2008).

5.1 Processing of samples and parasitological exams

Apart from two samples of sewage sludge was collected in all 100 samples of the produced compounds, with 25 before heat treatment and 75 after treatment (samples A, B and C). The samples were mixed one to one and 10 grams were weighed and stored with 30 ml of 10% formaldehyde solution for preservation of eggs, cysts and oocysts present (Duarte et al., 2008).

In the quantification of helminth eggs, protozoan cysts and oocysts, was used the technique of sedimentation of solids during 12 hours after the addition of filtered water to complete one litter. After this procedure, the method Bailenger modified by Ayres & Mara (1996), recommended by WHO was used. Two MacMaster chambers were prepared and for each sample were counted only viable eggs, obtaining the arithmetic mean of the two counts (Ayres & Mara, 1996).

Aiming to check the viability of eggs observed on microscopy was performed quantitative faecal culture technique described by Ueno, (1998). Two grams of the samples before heat treatment and after heat treatment were weighed (corresponding to B samples). The material was mixed with two grams of autoclaved dry sawdust and 10 ml of 1% sulfuric acid solution to fungi growth control. After 28 days of incubation at room temperature, was used the Baermann technique for collection and Keith (1953) key for the identification of larvae from samples obtained before and after heat treatment.

After counting of eggs and larvae of nematodes and protozoa cysts and oocysts from the 25 compounds analyzed, the data were transformed into log $(X +1)$ and submitted to the Student's t-test at 5% probability for comparison of the contamination before and after heat treatment.

5.2 Parasitological contamination of sewage sludge

In two s sewage sludge sample from Juramento city, North of Minas Gerais State of Brazil, the results revealed 1.2×10^4 eggs, 10^4 cysts and oocysts/kg for first sample and 2.7×10^4 eggs and 2.7×10^4 cysts and oocysts/ kg of the second sample, respectively. No infective larvae were recovered after quantitative faecal culture, indicating that the process of anaerobic digestion of sewage may have compromised the development and hatching of nematode larvae (Duarte et al., 2008).

In regions with hot weather, the utilization of up flow anaerobic sludge blanket (UASB) is a positive option for the treatment of domestic sewage. However, workers handling in these processes and the produced wastes are potential risk of being infected by parasites (Carvalho et al., 2003; Souza et al., 2006). The effects of aerobic and anaerobic digestion during 15 days on nematode eggs were evaluated by Black et al. (1982). The results indicated that 23% of *Ascaris* spp. eggs were destroyed in the anaerobic treatment and 38% with the aerobic. However, the anaerobic process had no effect on *Trichuris* spp. eggs and none of these treatments was effective for the *Toxocara* spp. eggs.

In a research in the region of Curitiba city,   Parana State of Brazil, the contamination of sewage sludge obtained in anaerobic fluidized sludge was evaluated. Were observed significant reductions in the number of viable eggs, present in the material from different treatment stations, ranging from 59.7 to 93%. The effectiveness of egg reduction was influenced by time and temperature processes and for seasonal effects (Paulino et al., 2001). In another study on the Paraiba State, Northeast of Brazil, was evaluated the effectiveness of the post-treatment. The three investigated systems were wetlands, rock beds and polishing ponds. The results indicated that the raw sewage and UASB effluents from the reactor had averages 353.7 and 50 eggs per litter, respectively. *Ascaris lumbricoides* prevailed on the other species in the raw sewage, with relative frequency of 56.5%. The three systems of post-treatment effluent produced wastes free of helminth eggs and this could be explained by sedimentation produced by a hydraulic surface of 0.20 m day-1 in ponds operated (Souza et al., 2005).

5.3 Parasitic contamination in the compounds before and after thermal treatment

In the table 2 are described the quantification of larvae recovered, counting helminth eggs and protozoan cysts and oocysts in organic compounds before and after heat treatment. These results demonstrate a high parasitic contamination of all the compounds analyzed and statistical analysis indicated no reduction of this contamination after heat treatment (Duarte et al., 2008). All samples were with levels above of one egg per 4 g of compost, which is not recommended for class A biosolids (Fernandes, 2000).

After the larval culture of the compound samples, before of the heat treatment, was observed that 52.4% of the larvae were of the genus *Cooperia*, 36.7% of the genus *Trichostrongylus*, 7.1% *Bunostomum* spp. and 3.8% to *Haemonchus* spp. After heat treatment of the compounds, 33.3% of the larvae were of the genus *Cooperia*, 30.7% of the *Trichostrongylus* spp., 2.7% to infective larvae of *Strongyloides* spp., which are commonly found as parasites of ruminants, and 33.3% of free living forms of the genus *Strongyloides* (Duarte et al., 2008).

The two samples of sewage sludge and the 100 samples of organic compounds showed high counts of viable helminth eggs under light microscopy. However, larvae of the cultures were recovered only in samples from the compounds and the genus identification suggested that the contamination came from the cattle manure or agricultural wastes contaminated with nematode eggs from ruminants (Duarte et al., 2008).

Compounds	eggs x 10^4/kg		larvaes x 10^4/kg		cysts e oocysts x 10^4/kg	
	before	after	before	after	before	after
avarage	33.9	22.3	25.3	41.9	52.8	66.6
standard deviation	31.3	21.2	23.6	71.5	54.7	82.3

(Duarte et al., 2008).

Table 2. Parasitic contamination in 25 different organic compounds produced with sewage sludge and agricultural residues before and after heat treatment at 60 ° C for 12 hours in North of Minas Gerais State of Brazil

Table 3 describes the average count of eggs, cysts and oocysts of protozoa from samples of three separate parcels (A, B and C) of each one of the 25 compounds, after the heat treatment at 60°C for 12 hours. There were no significant differences between samples of the same compound (P <0.05). The results revealed high counts of these parasitic structures in both the samples of the compounds after heat treatment, indicating that this process was not effective in the reducing of parasitic contamination in any of the sampled points (Duarte et al., 2008).

The data indicate that, even after composting and thermal treatment, the parasite eggs can remain viable and produce infective larvae. The compounds produced in this study could be classified as class B, for use on crops with little risk of contamination in relation to pathogenic organisms, such as cotton, orchards and plantations. These compounds also should not be applied to pasture for ruminants, since could be the source of contamination for gastrointestinal nematodes (Duarte et al., 2008).

The use of compounds with sewage waste for the fertilization of pastures should also not be displayed, because it could constitute a risk for contamination of the cattle with *Taenia saginata* eggs, contributing to the permanence of both cycles of the bovine cysticercosis and human taeniasis (Duarte et al., 2008).

compound Parcels	eggs x 10^4/kg			cyts and oocysts x 10^4/kg		
	A	B	C	A	B	C
averege	15.2	22.3	11.0	38.2	66.6	35.8
standard deviation	16	21.2	11.5	51.4	82.3	31.7

Duarte et al. (2008)

Table 3. Contamination with helminth eggs, cysts and oocysts of protozoan in three different samples of organic compounds produced from sewage sludge and agricultural residues after heat treatment at 60 ° C for 12 hours in the North of Minas Gerais, Brazil

6. Different waste treatments and parasitological contamination

Gaspard et al. (1995) also showed that nematode eggs are strongly resistant to most of the classical waste treatments. The work was carried out on sludges from various origins to properly evaluate the impact of the different treatments on nematode eggs. An extraction followed by a concentration procedure allowed isolation of eggs, the viability study being then performed on a culture. For the 19 samples with live eggs, all types of treatment were represented: fresh sludge, prolonged aeration, anaerobic digestion, lagooning, composting and liming. No egg development inhibition phenomenon was observed in fresh sludge. The test demonstrated there were 93% viable eggs. Biological treatments do not produce a total inactivation of nematode eggs. The samples submitted to such various treatments as prolonged aeration, anaerobic digestion, lagooning or composting still showed higher percentages of viable eggs. Prolonged aeration seemed to be totally ineffective with 93%

viable eggs recovered, and a lagoon containing 6-year-old sediments still showed 26% viable eggs. Regarding anaerobic digestion, 66% of viable eggs were recovered in the one sample. For the compost, the analysis on a small number of 8 eggs showed a viability of 25% and the chemical treatment with lime after 20 days of storage gave 66% of viable eggs.

Jhonson et al. (1998) evaluated an in-vitro test for the viability of *Ascaris suum* eggs exposed to various sewage treatment processes. After one week in a mesophilic anaerobic digester, 95% of *A. suum* eggs produced two-cell larvae in vitro, with 86% progressing to motile larvae. After five weeks in the digester, 51% progressed to motile larvae. Between 42% and 49% of eggs stored in a sludge lagoon for 29 weeks were viable and able to develop motile larvae. In the case of eggs that were embryonated before treatment, > 98% survived up to five weeks in the digester and were able to develop motile larvae. More than 90% of embryonated eggs survived for 29 weeks in the sludge lagoon and were able to develop motile larvae.

Solid waste landfill leachate and sewage sludge samples were quantitatively tested for viable *Enterocytozoon bieneusi, Encephalitozoon intestinalis, Encephalitozoon hellem,* and *Encephalitozoon cuniculi* spores by the multiplexed fluorescence in situ hybridization (FISH) assay. Depending on the variations utilized in the ultrasound disintegration, sonication reduced the load of human-virulent microsporidian (obligate intracellular parasites) spores to no detectable levels in 19 out of 27 samples (70.4%). Quicklime stabilization was 100% effective, whereas microwave energy disintegration was 100% ineffective against the spores of *E. bieneusi* and *E. intestinalis*. Top-soil stabilization treatment gradually reduced the load of both pathogens, consistent with the serial dilution of sewage sludge with the soil substrate. This study demonstrated that sewage sludge and landfill leachate contained high numbers of viable human-virulent microsporidian spores and that sonication and quicklime stabilization were the most effective treatments for the sanitization of sewage sludge and solid waste landfill leachates (Graczyk et al., 2007).

Kouja et al. (2010) assessed the presence and loads of parasites in 20 samples of raw, treated wastewater and sludge collected from six wastewater treatment plants. Samples were tested by microscopy using the modified Bailenger method (MBM), immunomagnetic separation (IMS) followed by immunofluorescent assay microscopy, and PCR and sequence analysis for the protozoa *Cryptosporidium* spp. and *Giardia* spp. The seven samples of raw wastewater had a high diversity of helminth and protozoa contamination. *Giardia* spp., *Entamoeba histolytica/dispar, Entamoeba coli, Ascaris* spp., *Enterobius vermicularis,* and *Taenia saginata* were detected by MBM, and protozoan loads were greater than helminth loads. *Cryptosporidium* sp. and *Giardia* sp. were also detected by IMS microscopy and PCR. Six of the eight samples of treated wastewater had parasites: helminths (n=1), *Cryptosporidium* sp. (n=1), *Giardia* sp. (n=4), and *Entamoeba* (n=4). Four of five samples of sludge had microscopically detectable parasites, and all had both genus *Cryptosporidium* sp. and *Giardia* sp. and its genotypes and subtypes were of both human and animal origin.

In another study evaluated the process of anaerobic digestion for treatment of cattle manure. After larvae cultures, positive results were obtained for the L3 larvae of *Haemonchus* spp., *Oesophagostomum* spp. and *Cooperia* spp. in the effluent, even after forty days of retention time (Amaral et al., 2004). However, Padilla & Furlong (1996) observed inactivating effect of anaerobiosis, close to 100%, with the retention time above of 56 days and according to Olson & Nansen (1987), mesophilic anaerobic digestion (35° C) and thermophilic (53° C) accelerated the inactivation of nematodes in relation to survival time of these parasites in conventional storage.

Sewage sludge and slurry are used as fertilizers on pastures grazed by ruminants. The interest of application on pastures of these two biowastes is environmental (optimal recycling of biowastes) and agronomic (fertilisation). The parasitic risk and the fertilisation value of such applications on pastures were evaluated during one grazing season. The sludge group of calves did not acquire live cysticerci and thus the risk was nil under the conditions of the study (delay of 6 weeks between application and grazing). The slurry group of calves did become lightly infected with digestive-tract nematodes, mostly *Ostertagia ostertagi*. Under the conditions of this experiment, a 6-week delay between application and grazing strongly reduced the risk of infection (Moussavou-boussougou et al., 2005a).

Helminth infection acquired by lambs grazing on pastures fertilised either by urban sewage sludge or cattle slurry were studied by Moussavou-boussougou et al. (2005b) in temperate Central Western France. The aim was to assess the risk of larval cestodoses in lambs after sewage application and of digestive tract nematode infection following the slurry application. The lambs did not acquire cysticercosis or any other larval cestodoses in the sewage sludge group and only very limited infections with *Cooperia* spp. and *Nematodirus* spp. were observed in the slurry group. It was concluded that the helminth risk was extremely low and was not a cause of restriction of the use of these biowastes.

7. Conclusion

The results obtained in the North of Minas Gerais, Brazil, showed that even after the composting of agricultural waste with sewage sludge and heat treatment at 60°C for 12 hours, large numbers of helminth eggs can remain viable. The use of the compounds with sewage sludge should be allocated to perennial crops and low risk of contamination for animals and humans is therefore not recommended for grazing ruminants, for horticulture or for the production of edible mushrooms.

The variation in data of other research to reduce parasitic contamination in composting and anaerobic digestion processes indicates the need for further research, standardizing and monitoring the waste to be recycled for agricultural or other purposes, to reduce risks to public health and animal infection. The initial contamination of sewage sludge used as well as time and temperature of the composting should be elucidated and the final compost produced should always be monitored as to risk of parasitic contamination that could be present.

8. References

Almeida, V.C. de; Lenzi, E.; Bortotti-Favero, L.O. & Luchese, E.B. (1998). Avaliação do teor de alguns metais e de nutrientes de lodos de estações de tratamento de Maringá. *Acta Scientiarum*, Vol.20, No.4, pp.419-425, Maringá, ISSN:1415-6814

Andreoli, C.V.; Fernandes, F. & Domaszak, S.C. (1997). Reciclagem agrícola do lodo de esgoto. pp. 81, Curitiba: SANEPAR

Andreoli, C. V. & Pegorini, E.S. (1998). Proposta de roteiro para elaboração de Planos de Distribuição de Lodo. In: *I Seminário sobre Gerenciamento de Biossólidos do Mercosul*, Curitiba

Andreoli, C.V.; Lara, A.I. & Fernandes, F. (1999). Reciclagem de biossólidos: transformando problemas em soluções. Sanepar; Finep, pp. 288, Curitiba

Amaral, C.V.; Amaral, L.A; Júnior, J.L.; Nascimento, A.A; Ferreira , D.S & Machado,M.R. (2004). Biodigestão anaeróbia de dejetos de bovinos leiteiros submetidos a diferentes tempos de retenção hidráulica. *Ciência Rural*, Vol. 36, pp.1897-1902, Santa Maria, ISSN: 0103-8478

Ayres R. & Mara, D. (1996). Analysis of wastewater for use inagriculture. *A laboratory manual of parasitological and bacteriological techniques*. pp. 31, Geneva, ISBN: 92 4 154484 8

Barreira, L.P.; Philippi Junior, A. & Rodrigues, M. S. (2006). Usinas de compostagem do Estado de São Paulo: qualidade dos compostos e processos de produção. *Eng. Sanit. Ambient.*, Vol. 11, No. 4, Rio de Janeiro, ISSN: 1413-4152

Bastos, R.K.X. & Mara, D.D. (1993). Avaliação de Critérios e Padrões de Qualidade Microbiológica de Esgotos Sanitários Tendo em Vista sua Utilização na Agricultura. *Proceedings of 17º Congresso Brasileiro de Engenharia Sanitária e Ambiental*. ABES, 19 a 23/09/93. Rio de Janeiro

Bonnet, B.R.P.; Lara, A.I. & Domaszak, S.C. (1998). Indicadores biológicos de qualidade sanitária do lodo de esgoto. In: *ANDREOLI, C.V.; BONNET, B.R.P. (Coord.) Manual de métodos para análises microbiológicas e parasitológicas em reciclagem agrícola de lodo de esgoto*. SANEPAR, pp.11-26, Curitiba

Black, M.I.; Scarpino,P.V.; O'donnell, C.J.; Meyer, K.B.; Jones, J.V. & Kaneshiro,E.S. (1982). Survival rates of parasite eggs in sludge during aerobic and anaerobic digestion. Applied Environmental Microbiology, Vol. 44, pp. 1138-1143, Washington, Online ISSN: 1098-5336, Print ISSN: 0099-2240

Carvalho, J.B.; Nascimento, E.R.; Neto, J.F.N.; Carvalho, I.S.; CarvalhO, L.S. & Carvalho J.S. (2003). Presença de ovos de helmintos em hortaliças fertilizadas com lodo de lagoa de estabilização. *Revista Brasileira de Análises Clínicas*, Vol. 35, pp.101-103, Rio de Janeiro

Chagas, W.F. (1999). Estudo de patógenos e metais em lodo digerido bruto e higienizado para fins agrícolas, das estações de tratamento de esgotos da Ilha do Governador e da Penha no estado do Rio de Janeiro. FIOCRUZ/ENSP, pp. 89, Dissertação (mestrado). Rio de Janeiro

Conselho de Meio Ambiente do Distrito Federal - CONAM/DF (2006). Resolução no 03/2006, de 18/7/2006. Diário Oficial do Distrito Federal no 138, de 20/7/2006, pp.10

Conselho Nacional do Meio Ambiente - CONAMA (2006). Resolução no 375/2006, de 29/8/2006. <http://www.mma.gov.br/port/conama/legiano/>. 29 Set. 2006

Crompton, D.W.T. (1999). How much human helminthiasis is there in the world? *Journal of Parasitology*, Vol. 85, pp. 379-403, Lawrence

David, A.C. (2002). Secagem térmica de lodos de esgoto: determinação da umidade de equilíbrio. 151f., Dissertação (Mestrado) – Escola Politécnica da Universidade de São Paulo, São Paulo

Doran, J.W. & Linn, D.M. (1979). Bacteriological quality of run off water from pastereland. Applied Microbiology, Vol. 37, pp. 985-991, Washington

Downey, N.E. & Moore, J.F. (1977). Trichostrongylid contamination of pasture fertilized with cattle slurry. Veterinary Record, Vol.101, No. 24, pp. 487-488, London, ISSN: 0042-4900

Duarte, E.R.; Almeida, A.C.; Cabral, B.L.; Abrão, F.O.; Oliveira, L.N.; Fonseca, M.P.& Arruda, R. (2008). Análise da contaminação parasitária em compostos orgânicos produzidos com biossólidos de esgoto doméstico e resíduos agropecuários. Ciência Rural, Vol. 38, pp.1279-1285, ISSN: 0103-8478

Environmental Protection Agency- EPA. (1995). A Guide to the Biosolids Risk Assessments for the EPA Part 503 Rule. S.l.:EPA 832-B-93-005. Sept. 1995. pp.144

Fernandes, F. (2000). Estabilização e higienização de biossólidos. In: BETTIOL, W.; CAMARGO, O.A. (Eds). Impacto ambiental do uso agrícola do lodo de esgoto. EMBRAPA Meio Ambiente, 2000. pp. 45-67. Jaguariúna

Furlong, J. & Padilha, T. (1996). Viabilidade de ovos de nematódeos gastrintestinais de bovinos após passagem em biodigestor anaeróbio. Ciência Rural, Vol. 26, pp. 269-271, Santa Maria, ISSN: 0103-8478

Gaspard, P.G.; Wiart, J. & Schwartzbrod, J. (1995). Urban sludge reuse in agriculture: waste treatment and parasitological risk. Bioresource Technology, Vol. 52, pp. 37-40, ISSN: 0960-8524

Graczyk, T. K.; Kacprzak, M.; Neczaj, E.; Tamang, L.; Graczyk, H.; Lucy, F.E. & Girouard, A.S. (2007). Human-virulent microsporidian spores in solid waste landfill leachate and sewage sludge, and effects of sanitization treatments on their inactivation. Parasitol Res., Vol. 101, pp. 569–575, print ISSN: 0932-0113, on line ISSN: 1432-1955

Instituto Ambiental do Paraná - IAP (2003). Instrução normativa para a reciclagem agrícola de lodo de esgoto, sem data. pp. 25

Johnson, P.W.; Dixon, R. & Ross, A.D. (1998). An in-vitro test for assessing the viability of Ascaris suum eggs exposed to various sewage treatment processes. International Journal for Parasitology, Vol. 28, pp. 627-633, ISSN: 0020-7519

Keith, R.K. (1953). The differentiation of the infective larvae of some common nematode. Australian Journal of Zoology, Vol. 1, pp. 223-235, Collingwood

Khouja, L.B.A.; Cama, V. & Xiao, L. (2010). Parasitic contamination in wastewater and sludge samples in Tunisia using three different detection techniques. Parasitol Res., Vol. 107, pp. 109–116, print ISSN: 0932-0113, on line ISSN: 1432-1955

Lopes, J.C; Ribeiro, L.G.; Araujo, M.G. & Beraldo, M.R.B.S. (2005). Produção de alface com doses de lodo de esgoto. Hortic. Bras. Vol. 23, No. 1, ISSN: 0102-0536

Olson, J.E. & Nansen, P. (1987). Inactivation of some parasites by anaerobic digestion of cattle slurry. Biological Wastes, Vol. 22, pp. 107-114, Fayetteville

Organización Mundial de la Salud – OMS (1989). Directrices Sanitárias sobre el uso de aguas residuales en agricultura y acuicultura. Série de Informes Técnicos, n.º 778. GINEBRA

Machado, M.F.S. (2001). A situação brasileira dos biossolidos. Dissertação (mestrado) - Universidade Estadual de Campinas . Faculdade de Engenharia Civil

Maria, I.C.; Kocssi, M.A.; Dechen, S.C.F. (2007). Agregação do solo em área que recebeu lodo de esgoto. Vol. 66, No. 2, pp. 291-298, Bragantia

Metcalf & Eddy (1991). Design of Facilities for the Treatment and disposal of Sludge. In: Wastewater engeneering - treatement, disposal and reuse. 3rd ed. U.S.A. McGraw-Hill International Editions, pp. 765-926

Moussavou-boussougou, M.N.; Geerts, S.; Madeline, M.; Ballandonne, C.; Barbier, D. & Cabaret, J. (2005a). Sewage sludge or cattle slurry as pasture fertilisers: comparative cysticercosis and trichostrongylosis risk for grazing cattle. Parasitol Res., Vol. 97, pp. 27-32, print ISSN: 0932-0113, on line ISSN: 1432-1955

Moussavou-Boussougou, M.N.; Dorny, P. & Cabaret, J. (2005b). Very low helminth infection in sheep grazed on pastures fertilised by sewage sludge or cattle slurry. Veterinary Parasitology, Vol 131, pp. 65-70, ISSN: 0304-4017

Paulino, R.C.; Castro, E.A.; Thomaz-Soccol, V. (2001). Tratamento anaeróbio de esgoto e sua eficiência na redução da viabilidade de ovos de helmintos. Revista da Sociedade Brasileira de Medicina Tropical, Vol. 34, pp. 421-428, Uberaba, ISSN 0037-8682

Roque, O.C.C. (1997). Sistemas Alternativas de Tratamento de Esgotos Aplicáveis as Condições Brasileiras – Tese de Doutorado em Saúde Pública, , pp. 153, FIOCRUZ – Rio de Janeiro

Companhia de Saneamento do Paraná – SANEPAR (1997). Manual Técnico para Utilização Agrícola do lodo de esgoto no Paraná, pp. 96

Sousa, J.T.; Ceballos, B.S. O.; Henrique, I. N.; Dantas, J.P. & Lima, S.M.S.. (2006). Reuso de água residuária na produção de pimentão (Capsicum annuum L.). Revista Brasileira de Engenharia Agrícola e Ambiental, Vol. 10, pp. 89-96, Campina Grande, ISSN: 1415-4366

Sousa, J. T. & Leite, V. D. (2005). Tratamento de esgoto para uso na agricultura do Semi-Árido Nordestino. Engenharia Sanitária Ambiental, Vol. 10, pp. 260-265, Rio de Janeiro, ISSN : 1413-4152

Tedesco, M.J.; Gianello, C.; Bissani, C.A.; Bohnen, H. & Volkweiss, S.J. (1995). Análise de solo, plantas e outros materiais. 2.ed. Porto Alegre: Departamento de Solos/UFRGS, pp. 174

Thomaz-Soccol, V. (1998). Aspectos sanitários do lodo de esgoto. In: Seminário sobre gerenciamento de biossólidos do mercosul, 1., Anais... Curitiba: SANEPAR/ABES, pp. 65-75, Curitiba

Ueno, H. & Gonçalves, P.C. (1998). Manual para diagnóstico das helmintoses de ruminantes. 4.ed. Tokio: Japan International Cooperation Agency, pp. 143

United States Environmental Protection Agency – USEPA (1995). A guide to the biosolids risk assessments for the EPA Part 503 rule, 1995. Washington: Office of Wastewater Management, EPA/832-B-93-005, pp. 195

Veras, L.R. V. & Povinelli, J. (2004). A vermicompostagem do lodo de lagoas de tratamento de efluentes industriais consorciada com composto de lixo urbano. *Engenharia Saniária Ambiental*, Vol. 9, No. 3, Rio de Janeiro, ISSN : 1413-4152

World Health Organization. (1989). Health guidelines for the use of wastewater in agriculture and aquaculture. Technical report series. 778. Geneva: WHO, pp. 72

2

Heavy Metal Contamination of Zn, Cu, Ni and Pb in Soil and Leaf of *Robinia pseudoacacia* Irrigated with Municipal Wastewater in Iran

Masoud Tabari[1]*, Azadeh Salehi[1],
Jahangard Mohammadi[2] and Alireza Aliarab[3]
[1]*Tarbiat Modares University,*
[2]*Shahrekord University*
[3]*Gorgan University,*
Iran

1. Introduction

The economic development of the society towards large-scale urbanization and industrialization is leading to production of huge quantities of wastewaters (Singh & Agrawal, 2008). Wastewaters can be used for the restoration of degraded land (Madejo´n et al., 2006), and the growth of vegetation having commercial and environmental value (Aggeli et al., 2009). Establishment of tree plantations following wastewater irrigation has been a common practice for many years (Kalavrouziotis & Arslan-Alaton, 2008). Several researches of wastewater irrigated plantations in many countries such as India (Bhati & Singh, 2003; Singh & Bhati, 2005), Australia (Sharma & Ashwath, 2006), New Zealand (Guo et al., 2002; Kimberley et al., 2003), Sweden (Hasselgren, 2000), Canada (Cogliastro et al., 2001), Hungary (Vermes, 2002), etc. are available.

In Iran, huge section of useful water of major metropolitan cities converts to the municipal wastewater (Tajrishi, 1998). Since the deficiency of access to adequate water for irrigation is a matter of increasing concern and limiting factor to develop plantation, therefore municipal wastewater could be utilized as an important source of water for expansion of tree plantations in and around the city and industrial complexes (Al-Jamal et al., 2000; Kalavrouziotis & Apostolopoulos, 2007; Salehi et al., 2007). This practice not only reduces the toxicity of soil and plays an important role in safeguarding the environment, because woody species may utilize wastewater and uptake heavy metals through extensive root systems and retain them for a long time (Madejo´n et al., 2006), but it also creates opportunities for commercial biomass production and sequestration of excess minerals in the plant system (Sharma & Ashwath, 2006).

Again, wastewaters may contain amounts of potentially harmful components such as heavy metals and pathogens (Rattan et al., 2005; Toze, 2006). The effects of microbial pathogens are usually short term and vary in severity depending on the potential for human, animal or environmental contact (Toze, 2006), while the heavy metals have longer term impacts that could be a source of contamination and be toxic to the soil (Sharma et al., 2007) and plant (Gasco´ & Lobo, 2007). Hence if wastewater is to be recycled safely for irrigation, the problems associated with using it need to be known (Emongor & Ramolemana, 2004).

According to differences in climatic, vegetation, socio-economic conditions and also in quality of soil and wastewater between different regions and even within different time periods in one region, utilizing only the applicable guidelines to other regions of the world would be a mistake and in long-term would damage the soil and water resources, therefore local researches need to be carried out (Kalavrouziotis & Arslan-Alaton, 2008).

Robinia pseudoacacia L. (black locust) is native to the southeastern United States, but has been widely planted and naturalized elsewhere. *R. pseudoacacia* trees have nitrogen- fixing bacteria on its root system, for this reason it can grow on poor soils, therefore it can improve fertility of soil. In Iran it often planted alongside streets, in green space and parks, especially in large cities, because it tolerates pollution well (Mossadegh, 1993). The use of municipal wastewater in growing *R. pseudoacacia* in suburban areas could be beneficial for the economic disposal of wastewater, defers ecological degradation by containing the pollutants in the soil and growth of vegetation having aesthetic and environmental value. The present study was carried out around Tehran, Iran, where wastewater has been commonly used for irrigation of peri-urban crops for many years. The objective of the present report is to quantify concentration and contamination of Zn, Cu, Ni and Pb in irrigation water, soil and leaves of *R. pseudoacacia* trees from site having long-term use of wastewater for irrigation of land.

2. Materials and methods

2.1 Site description

The study site is located in Shahr-e Rey, 5 Km south of Tehran-Iran (Latitude 35° 37' N, Longitude 51° 23' E, 1005 m above sea level). The climate of the area is semi-arid with mild-cold winters and 7 months (Mid April-Mid November) dry season. Average annual rainfall and average annual temperature are 232 mm and 13.3° C, respectively. The highest rainfall is in March (41.32 mm) and the lowest in August (0.89 mm). The warmest month occurs in August and the coldest in January.

Experiments were conducted at two even-aged (15 years) artificial stands of black locust in October 2006. The first stand was irrigated with municipal wastewater and the second with well water since they were planted. Durations of irrigation were based on tree water-use and the potential evapo-transpiration, which varied seasonally in response to the climate and on an average the irrigations were carried on 8 day durations for 8 months/year (April-November). The soil of two stands were both clay-loam (according to US soil taxonomy) with 29.25% clay, 36.20% silt and 34.55% sand in the stand irrigated with municipal wastewater and 27.14% clay, 37.86% silt and 35% sand with well water (Table 2).

2.2 Plant and soil sampling

For the sampling of leaf and soil, four plots were randomly identified in each stand. Plots were 30 m × 30 m, with tree spacing of 3 m × 4 m. In each plot, four trees were selected and in the growing season leaf samples of *Robinia pseudoacacia* trees taken from the top of crown and the part affected by sunlight (Habibi Kaseb, 1992). This collection provided 16 leaf samples in each stand. At the end of the sampling, one representative leaf sample from each plot (by mixing of four samples of each plot) was taken (decreasing of sample quantity for chemical analysis). Soil samples were taken under each selected tree from the root zone at a depth interval of 15 cm down to 60 cm by digging profiles. This collection provided 48 soil samples in each stand from three depths (0-15, 15-30 and 30-60 cm). At the end of soil sampling, three representative soil samples of three depths from each plot were taken by mixing of samples of each layer in each plot (decreasing of sample quantity for chemical analysis) according to Habibi Kaseb (1992). Municipal wastewater and well water were sampled daily (3 days in each month) from early June to late November at three-hour intervals (7 am, 13 pm and 19 pm) to make a composite sample of each day.

2.3 Laboratory analysis

Concentrated HNO_3 was added to the water samples to avoid microbial utilization of heavy metals (Sharma et al., 2007) and then they were brought to the laboratory in resistant plastic bottles to avoid adherence to the container wall. They were filtered through a Whatmann 42 mm filter paper and stored at 4 °C to minimize microbial decomposition of solids (Yadav et al., 2002; Bhati & Singh, 2003). Some parameters were measured separately, pH and EC by the procedure described using OMA (1990) and heavy metals (Zn, Cu, Ni and Pb) of water samples were estimated by the aqua regia method of Jackson (1973) followed by a measurement of concentrations using an Atomic Absorption Spectrophotometer (model-3110, Perkin-Elmer, Boesch, Huenenberg, Switzerland).

The soil samples air-dried, crushed, passed through a 2 mm sieve and were analyzed for various physico-chemical properties. Soil texture was determined using the hydrometer method according to Bouyoucos (1965). Soil pH and electrical conductivity (EC) were determined in 1:2 soil:water suspension by pH and EC meters (Hati et al., 2007). Soil organic carbon (SOC) content was determined by the Walkley–Black method (Nelson & Sommers, 1996). Calcium carbonate ($CaCO_3$) was measured with a calcimeter. The concentration of soil heavy metals (Zn, Cu, Ni and Pb) was extracted after digestion with 3:1 concentrated HCl–HNO_3 and measured by Atomic Absorption Spectrophotometer (Gasco´ & Lobo, 2007).

Leaf samples were washed using tap water, rinsed with distilled water, oven dried at 80 °C for 24 h, ground in a stainless steel mill and retained for chemical analysis (Singh & Bhati, 2005). For determination of heavy metal concentration (Zn, Cu, Ni and Pb), the leaf samples were wet digested as per Jackson (1973) and were measured using an Atomic Absorption Spectrophotometer.

2.4 Statistical analysis

Average leaf heavy metals and soil physico-chemical properties of two stands (irrigated with municipal wastewater and irrigated with well water), compared using independent-samples t-test (Pelosi & Sandifer, 2003). Data of soil heavy metals were analyzed for differences due to depth in the profile using one-way ANOVA. Furthermore, the variations in EC, pH and heavy metals of municipal wastewater and well water were also tested using independent-samples t-test. All the data were analyzed using the SPSS statistical package (Lindaman, 1992).

3. Results and discussion

3.1 Physico-chemical properties of wastewater and well water

The quality of municipal wastewater and well water was assessed for irrigation with respect to their pH, EC, and concentration of heavy metals (Table 1). Results indicated that the waters were alkaline in reaction. The pH of the municipal wastewater in various months ranged from 7.51 to 7.75 and 6.69 to 7.62 for well water. The EC of wastewater ranged from 1.78 to 2.12 dS/m with the greatest value detected in August. The average EC of municipal wastewater exceeded 1 dS/m (1.91 dS/m) indicating that this wastewater was saline in nature (Rattan et al., 2005). The pH and EC of municipal wastewater were significantly (P < 0.01) higher than the well water. The concentration of heavy metals (Zn, Cu, Pb and Ni) tended to be higher in municipal wastewater. In water samples, Zn, Cu, Pb and Ni concentrations were 0.43, 0.09, 0.033 and 0.028 mg/l, respectively in well water samples, whereas, corresponding values for wastewater were 3.30, 1.26, 0.106 and 0.081 mg/l. On an average, wastewater contained 7.67, 14, 3.21 and 2.89 times higher amounts of Zn, Cu, Pb and Ni respectively compared to well water. The comparison of measured factors with WHO (World Health Organization) standard showed that water used for irrigation based on pH and EC were in a normal range, however based on heavy metals: Pb and Ni concentration of municipal wastewater and well water was higher than standard range. Zn concentration of municipal wastewater also was higher than the standard but Cu concentration was normal. The concentration of these two elements was lower than the standard in well water (Table 1).

Parameters	Units	Municipal wastewater	Well water	WHO*
pH	----------	7.63 ± 0.01 [a]	7.32 ± 0.05 [b]	6.5 - 8.5
EC	(dS/m)	1.91 ± 0.02 [a]	0.590 ± 0.008 [b]	3
Zn	(mg/l)	3.30 ± 0.06 [a]	0.43 ± 0.07 [b]	3
Cu	(mg/l)	1.26 ± 0.03 [a]	0.09 ± 0.01 [b]	1-2
Pb	(mg/l)	0.106 ± 0.063 [a]	0.033 ± 0.026 [b]	0.01
Ni	(mg/l)	0.081 ± 0.007 [a]	0.028 ± 0.002 [b]	0.02
Different superscripts in row indicate significant (P < 0.01) difference. Values are mean of eighteen replications (3 days * 6 months) with ± SE, * Hach, 2002				

Table 1. Characteristics of municipal wastewater and well water

3.2 Impact of municipal wastewater irrigation on soil properties

Data of Table 2 indicate that application of municipal wastewater were resulted an increase (0-60 cm soil layer; mean of soil layers) in pH, EC, C, organic matter and $CaCO_3$ of wastewater-irrigated soil as compared to well water-irrigated soil. Increase in pH was 1.02 unit and EC 1.68 times in soil of wastewater treatment compared to the soil of well water treatment. The increase in pH and EC of soil in the wastewater-irrigated stand may have been due to alkaline nature of municipal wastewater (Singh & Bhati, 2005). SOC as a basic index of soil playing a variety of roles in nutrient, water, and biological cycles (Rattan et al., 2005) was 1.17%–1.29% in municipal wastewater-irrigated soil, whereas it was 0.88%–1.14%

Soil properties	Clay (%)	Silt (%)	Sand (%)	texture	pH	EC (dS/m)	C (%)	Organic matter (%)	CaCO₃ (%)
Wastewater irrigated soil	29.25	36.20	34.55	Clay loam	8.17 [a] (0.03)	1.28 [a] (0.04)	0.718 [a] (0.032)	1.23 [a] (0.05)	20.20 [a] (0.57)
Well water irrigated soil	27.14	37.86	35	Clay loam	7.94 [b] (0.10)	0.763 [b] (0.036)	0.585 [b] (0.062)	1.00 [b] (0.107)	18.55 [b] (0.45)
Values are mean of four replications with ± SD in parentheses; Different superscripts in columns indicate significant (P < 0.01) difference									

Table 2. Soil properties of two stands (0-60 cm)

in soil irrigated with well water. Increase in SOC content might be due to municipal wastewater application (Bhati & Singh, 2003). In general, the suitability of soils for receiving wastewater without deterioration varies widely, depending on their infiltration capacity, permeability, cation exchange capacities, phosphorus adsorption capacity, texture, structure, and type of clay mineral (Ivan & Earl, 1972).

The concentration of heavy metals (Zn, Cu, Pb and Ni) was higher in all depths of wastewater irrigated soil compared to those of well water irrigated soil (Fig. 1). As a matter of fact, high concentration of heavy metals in wastewater leads to increase them in soil (Huerta et al., 2002; Nan et al., 2002; Mapanda, et al., 2005). The comparison of soil Zn, Cu, Pb and Ni with critical range of heavy metals in soil (Table 3) showed that only Ni of soil treated with municipal wastewater and Pb of soil treated with the both municipal wastewater and well water were higher than the standard amounts of soil. The effects of wastewater irrigation on accumulation of soil heavy metals depend on various factors such as concentration of wastewater heavy metals, the period of wastewater irrigation, and soil properties (pH, texture, organic matter) (Rattan et al., 2005). And also generally, 10 to 50 years is needed so that the heavy metal levels precede the standard levels (Smith et al., 1996). Because of the high concentration of Pb in all soil and water samples, it can be predicted that besides the municipal wastewater, Pb probably has been added to the water and soil from other sources such as air pollution.

In the present investigation the concentration of heavy metals decreased with soil depth in both stands (Fig. 2). These results are in agreement with the findings obtained later (Yadav

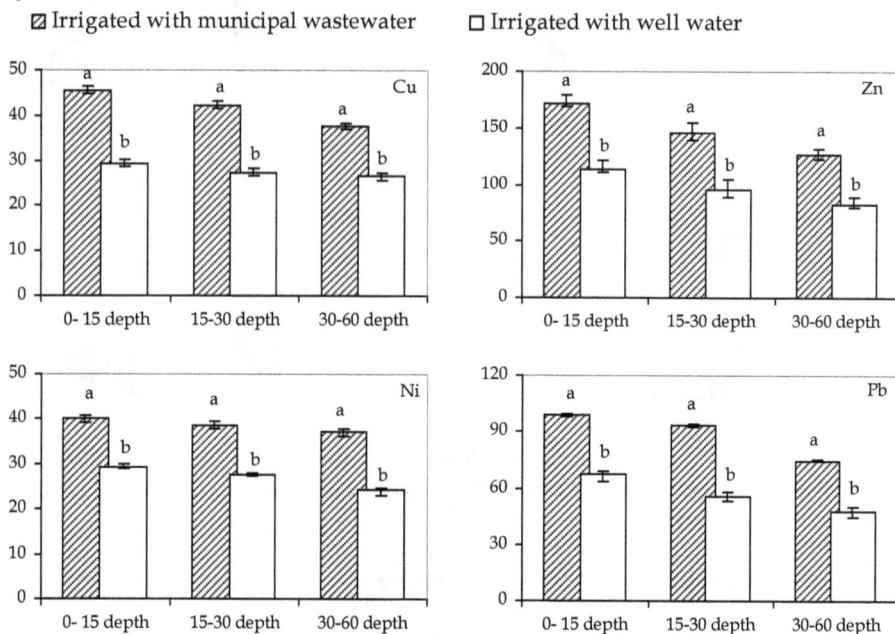

Fig. 1. Comparison of heavy metals in similar depths (0-15, 15-30 and 30-60 cm) between soils irrigated with wastewater and well water (mg/kg); Error bars are ± SE

et al., 2002). Since, the soil surface is richer in heavy metals than the underlying layers, greater accumulation in the topsoil probably is due to soil texture (the soil texture in both stands is clay-loam, as a result penetrability is decreased and accumulation of heavy metals are often observed at upper layers), low mobility of heavy metals in soil (Afyoni et al., 1998), and surface application of municipal wastewater.

Heavy metals	Critical range *
Zn (mg/kg)	10-500
Cu (mg/kg)	5-400
Pb (mg/kg)	40
Ni (mg/kg)	30
* Zn and Cu: Salardiny (1992); Pb and Ni: (EPA)	

Table 3. Critical range of heavy metals in soil

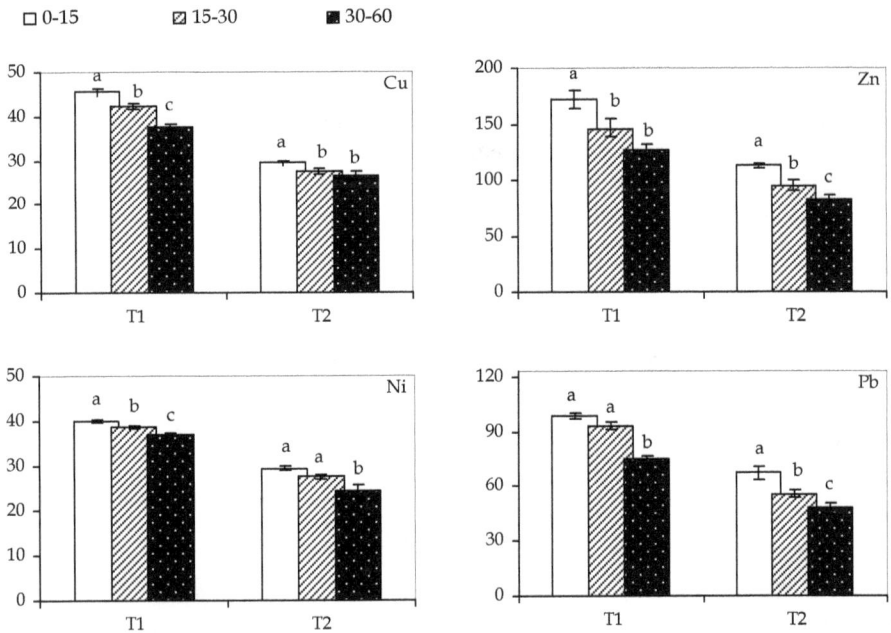

Fig. 2. Comparison of heavy metals among different depths (0-15, 15-30 and 30-60 cm) of soil in each irrigated stand (mg/kg); T_1: Soil irrigated with wastewater, T_2: Soil irrigated with well water; Error bars are ± SE

3.3 Changes in concentration of leaf heavy metals
The concentration of Zn and Cu elements in the leaves of black locust trees differed significantly under impact of two irrigation treatments. These concentrations in the leaves of wastewater irrigated trees were about 1.5 times higher than those of well water irrigation. However, irrigation with municipal wastewater did not result in toxicity to Zn and Cu of leaves (Table 4). Marked difference in Zn and Cu of tree leaves may be due to the increase of

them through municipal wastewater (Meli et al., 2002). This result is in agreement with Singh & Bhati (2005) and Aghabarati et al. (2008), where substantially greater concentration of these elements were observed in leaf of *Dalbergia sissoo* seedlings and *O. europaea* trees irrigated with municipal wastewater compared to control. Ni and Pb were not detected in leaf samples which may be due to the low dynamic of heavy toxic metals, whereas it was likely accumulated in lower parts of the plant, such as root and stem. Nevertheless, Madejo´on et al. (2006) reported the presence of some heavy toxic metals in leaf of olive and holm oak trees. In fact, the quantity of element absorption using plant depends upon many factors including the total quantity of the elements applied through wastewater application, soil properties, and type of plant (Bozkurt & Yarilga, 2003; Kalavrouziotis and Arslan-Alaton, 2008).

Heavy metals	Wastewater	Well water	P-value	Range*
Zn (mg/kg)	30.62 ± 6.00 [a]	20.63 ± 2.60 [b]	<0.05	10-100
Cu (mg/kg)	4.87 ± 0.77 [a]	2.81 ± 0.23 [b]	<0.01	2-20
Ni (mg/kg)	nd	nd	------	--------
Pb (mg/kg)	nd	nd	------	--------
Values are mean of four replications with ± SD; different superscripts in rows indicate significant difference; nd: not detected; * Salardiny (1992)				

Table 4. Concentration of heavy metals in leaf of black locust trees irrigated with wastewater and well water

4. Conclusion

Today, the reuse of municipal wastewater for land irrigation constitutes a practical method of disposal which is expected to contribute decisively to the handling and minimization of environmental problems arising from the disposal of wastewater effluents on land and into aquatic systems. The application of wastewaters onto appropriate forest species will enable long term environmental protection, creating a new water source in significant quantities for the irrigation of forested areas at the same time. Again, the use of wastewaters for irrigating maybe increases heavy metals and pathogens in soil and plant. Hence, the control of all of parameters associated with the disposal of wastewaters on land should be done for safe reuse of them. Furthermore, the method and extent of use of wastewaters, however, vary according to the infrastructure and the local socio-economic conditions prevalent from country to country.

According to the results of the present paper from the area under study where municipal wastewater is being used for about 15 years, high level of some heavy metals in irrigation water and soil treated with municipal wastewater and possibility of accumulation of heavy toxic metals in lower parts of the plant, it is said that regulations about the utilization of municipal wastewater in irrigation should consider in order to minimize the risk of negative effects to ecosystem health. This can be controlled by avoiding toxic elements from entering the municipal wastewater and continued monitoring or treatment of wastewaters before it is put into disposal channel for irrigation.

5. Acknowledgments

This project was supported by the Tarbiat Modares University Research Fund. Authors are grateful to Forestry Department of the Tarbiat Modares University for technical and scientific assistance and Shahr-e-Ray Municipality for its support on field assistance of this research.

6. References

Afyoni, M.; Rezainejad, Y. & Khayyambashi, B. (1998). Effect of sewage effluent on function and absorb of heavy metals by spinach and lettuce. *Journal of Science and Technology of Agriculture and Natural Resources*, 2, 1, 19-30

Aggeli, K.; Kalavrouziotis, I.K. & Bezergianni, S. (2009). A proposal of a treated wastewater reuse design system in urban areas. *Fresenius Environmental Bulletin*, 18, 7b, 1295-1301

Aghabarati, A.; Hosseini, S.M.; Esmaili, A. & Maralian, H. (2008). Growth and mineral accumulation in *Olea europaea L.* trees irrigated with municipal effluent. *Research Journal of Environmental Sciences*, 2, 4, 281-290

Al-Jamal, M.S.; Sammis, T.W.; Mexal, J.G.; Picchioni, G.A. & Zachritz, W.H. (2000). A growth_irrigation scheduling model for waste water use in forest production. *Agriculture Water Management*, 56, 1, 57-79

Bhati, M. & Singh, G. (2003). Growth and mineral accumulation in *Eucalyptus camaldulensis* seedlings irrigated with mixed industrial effluents. *Bioresource Technology*, 88, 3, 221-228

Bozkurt, M.A. & Yarilga, T. (2003). The effects of sewage sludge applications on the yield, growth, nutrition and heavy metal accumulation in apple trees growing in dry conditions. *Turkish Journal of Agriculture & Forestry*, 27, 285-292

Bouyoucos, G.J. (1965). Hydrometer method for making particle size analysis of soils. *Agronomy Journal*, 54, 464-465

Cogliastro, A.; Gerald, D. & Daigle, S. (2001). Effects of wastewater sludge and woodchip combinations on soil properties and growth of planted hardwood trees and willows on a restored site. *Ecological Engineering*, 16, 4, 471-485

Emongor, V.E. & Ramolemana, G.M. (2004). Treated sewage effluent (water) potential to be used for horticultural production in Botswana. *Physics and Chemistry of the Earth*, 29, 15-18, 1101-1108

EPA: www.EPA.org

Gasco´, G. & Lobo, M.C. (2007). Composition of Spanish sewage sludge and effects on treated soil and olive trees. *Waste Management*, 27, 11, 1494-1500

Guo, L.B.; Sims, R.E.H. & Horne, D.J. (2002). Biomass production and nutrient cycling in eucalyptus short rotation energy forests in New Zealand. I: biomass and nutrient accumulation. *Bioresource Technology*, 85, 3, 273-283

Habibi Kaseb, H. (1992). *Forest Pedology*. Tehran University Publications, Tehran, Iran

Hach, C. (2002). *Water Analysis Handbook*. Loveland, Colorado, USA. WHO, 61-62

Hasselgren, K. (2000). Use of municipal waste products in energy forestry: highlights from 15 years of experience. *Biomass and Bioenergy*, 15, 1, 71-74

Hati, K.M.; Biswas, A.K.; Bandyopadhyay, K.K. & Misra, A.K. (2007). Soil properties and crop yields on a vertisol in India with application of distillery effluent. *Soil Tillage research*, 92, 1-2, 60-68

Huerta, L.; Contreras-Valadez, R.; Palacios-Mayorgac, S.; Miranda, J. & Calva-Vasque, G. (2002). Total elemental composition of soils contaminated with waste water irrigation by combining IBA techniques. *Nuclear Instruments & Methods in Physics Research Section B-Beam Interactio*, 189, 1-4, 158-162

Ivan, F.S. & Earl, E.A. (1972). *Soil limitations for disposal of municipal waste waters*. Michigan State University Research Report, 195: 54

Jackson, M.L. (1973). *Soil chemical analysis*. Prentice Hall of India Private Ltd, New Delhi

Kalavrouziotis, I.K. & Apostolopoulos, C.A. (2007). An integrated environmental plan for the reuse of treated waste water effluents from WWTP in urban areas. *Building and Environment*, 42, 4, 1862-1868

Kalavrouziotis, I.K. & Arslan-Alaton, I. (2008). Reuse of urban wastewater and sewage sludge in the Mediterranean countries: case studies from Greece and Turkey. *Fresenius Environmental Bulletin*, 17 ,6, 625-639

Kimberley, O.M.; Wang, H.; Wilks, J.P.; Fisher, R.C. & Magesan, N.G. (2003). Economic analysis of growth response from a pine plantation forest applied with biosolids. *Forest Ecology and Management*, 189, 1-3, 345-351

Lindaman, H.R. (1992). *Analysis of variance in experimental Design*. Springer-Verlag, New York

Madejo´n, P.; Maran˜o´n, T. & Murillo, J.M. (2006). Biomonitoring of trace elements in the leaves and fruits of wild olive and holm oak trees. *Science of the Total Environment*, 355, 1-3, 187-203

Mapanda, F.; Mangwayana, E.N.; Nyamangara, J. & Giller, K.E. (2005). The effect of long-term irrigation using waste water on heavy metal contents of soils under vegetables in Harare, Zimbabwe. *Agriculture Ecosystems & Environment*, 107, 2-3, 151-165

Meli, S.; Porto, M.; Belligno, A.; Bufo, S.A.; Mazzatura, A. & Scopa, A. (2002). Influence of irrigation with lagooned urban wastewater on chemical and microbiological soil parameters in a citrus orchard under Mediterranean condition. *Science of the Total Environment*, 285, 1-3, 69-77

Mossadegh, A. (1993). *Afforestation and Forest Nursery*. Tehran University Publications, Tehran, Iran

Nan, Z.Li.; Zhang, J. & Cheng, G. (2002). Cadmium and zinc interaction and their transfer in soil-Crop system under actual field conditions. *Science of the Total Environment*, 285, 1-3, 187-195

Nelson, D.W. & Sommers, L.E. (1996). *Total carbon, organic carbon and organic matter*. In: Bigham, J.M. (Ed.), Methods of Soil Analysis: Part3. Chemical Methods. SSSA, Madison, 961-1010

OMA, (1990). *Official methods of analysis, 15th ed*. Association of Official Analytical Chemists, Arlington, Virginia, USA

Pelosi, M.K. & Sandifer, T.M. (2003). *Elementary statistics: from Discovery to Decision* ‹John Willey & Sons, INC. 793 pp

Rattan, R.K.; Datta, S.P.; Chhonkar, P.K.; Suribabu, K. & Singh, A.K. (2005). Long-term impact of irrigation with sewage effluents on heavy metal content in soils, crops and groundwater-a case study. *Agriculture Ecosystems & Environment*, 109, 3-4, 310-322

Salardiny, A. (1992). *Soil fertility*. Tehran University Publications, Tehran, Iran

Salehi, A.; Tabari, M.; Mohammadi, J. & Ali-Arab, A.R. (2007). Growth of black locust irrigated with municipal effluent in green space of southern Tehran. *Research Journal of Environmental Sciences* , 1, 5, 237-243

Sharma, A. & Ashwath, N. (2006). Land disposal of municipal effluents: Importance of choosing agroforestry systems. *Desalination*, 187, 1-3, 361-374

Sharma, R.K.; Agrawal, M. & Marshall, F. (2007). Heavy metal contamination of soil and vegetables in suburban areas of Varanasi, India. *Ecotoxicology and Environmental Safety*, 66, 2, 258-266

Singh, G. & Bhati, M. (2005). Growth of *Dalbergia sissoo* in desert regions of western India using municipal effluent and the subsequent changes in soil and plant chemistry. *Bioresource Technology*, 96, 9, 1019-1028

Singh, R.P. & Agrawal, M. (2008). Potential benefits and risks of land application of sewage sludge. *Waste Management*, 28, 2, 347-358

Smith, C.J.; Hopmans, P. & Cook, F.J. (1996). Accumulation of Cr, Pb, Cu, Ni, Zn and Cd in soil following irrigation with treated urban effluent in Australia. *Environmental Pollution*, 94, 3, 317-323

Tajrishi, M. (1998). New and comprehensive outlook to the problem of municipal effluent of Tehran. *Journal of Water & Effluent*, 28, 16-30

Toze, S. (2006). Reuse of effluent water-benefits and risks. *Agriculture Water Management*, 80, 1-3, 147-159

Vermes, L. (2002). Poplar plantations for waste water treatment and utilization in Hungary. *IWA Regional Symposium on Water Recycling in Mediterranean Region*, pp. 297-300, Iraklio, Greece, 26-29 September

Yadav, R.K.; Goyal, B.; Sharma, R.K.; Dubey, S.K. & Minhas, P.S. (2002). Post-Irrigation impact of domestic sewage effluent on composition of soils, crops and ground water-a case study. *Environment International*, 28, 6, 481-486

3

Impact of Municipal Wastewater on Growth and Nutrition of Afforested *Pinus eldarica* Stands

Masoud Tabari, Azadeh Salehi and Jhangard Mohammadi
Tarbiat Modares University
Iran

1. Introduction

As a whole, water is a most important source for plantations particularly in the dry regions (Mosadegh, 1999). In other hand, wastewater can be used to cover the needs of urban and rural areas and parks as well as industrial complexes to develop green space and to reduce air pollution (Al-Jamal *et al.*, 2000; Singh and Bhati, 2005; Sharma, *et al.*, 2007). In reality, wastewater except the water resource for irrigating the plantations is an enormous nutrient source, too (Meli *et al.*, 2002; Rattan *et al.*, 2005). Of course, establishment of trees plantation for waste water irrigation has been a common practice for many years. The practice not only defers ecological degradation by the pollutants in the soil, because trees are long-living organisms which can take up trace elements from the soil, water or air and retain them for a long time (Madejo´n *et al.*, 2006). But it also creates opportunities for commercial biomass production and sequestration of excess minerals in the plant system (Sharma and Ashwath, 2006). Therefore, the use of waste water in growing woodlots is a viable option for the economic disposal of waste water (Neilson *et al.*, 1989). Moreover, waste water from municipal origin is rich in organic matter and also contains appreciable amounts of macro and micro-nutrients (Gupta *et al.*, 1998). Accordingly nutrients levels of soil are expected to improve considerably using continuous irrigation with municipal waste water (Ramirez-Fuentes *et al.*, 2002; Rattan, *et al.*, 2005). Apart from this, in the case of the utilization of wastewater mixed with harmful heavy metals lead to decrease the toxicity, through a developed rooting system in plantations (Karpiscak *et al.*, 1996) and as such, play the important and fundamental role for the environmental protection (Cromer *et al.*, 1987; Stewart *et al.*, 1990). However, this can not be ignored that the use of wastewater for irrigation purposes might damage the ecosystem because the high toxic concentration and heavy metals (Gupta *et al.*, 1998; Brar *et al.*, 2000; Yadav *et al.*, 2002). The accumulation of heavy metals in soil is related to pH, texture and cation exchange capacity of soil (Datta *et al.*, 2000). Therefore, decision about the application of wastewater should be made based on the views of specialties of water, soil, plant and environment of every location (Nagshinepour, 1998).

Iran is a part of arid regions in the world being encountered acute crises owing to the increased population and need of water resources (Tabatabaei, 1998). It is noteworthy saying that thousands liters of domestic, industrial and hospital effluents are daily flowing from Tehran metropolitan area and influence the underground water resources. In the same way, 80 percent of the useful water of the citizens in Tehran is also transformed as municipal effluent (Tajrishi, 1998). On the other side, unplanned expansion and air pollution of Tehran make it unavoidable to increase the green space. In reality, urban green space and green belt around the city can play an effective role in air purification and climate health. Since the lack of water is a limiting factor for development of green space, therefore municipal effluent may be suitable (Torabian and Hashemi, 1999).

Till now inside the country several researches have been conducted about effect of municipal effluent on soil and agricultural crops, but not on softwoods. The objective of this study was to investigate the effects of the 15 years municipal waste water application on the growth of *Pinus eldarica* Medw. trees and the minerals accumulation in the trees needles.

2. Materials and methods

The study site is an abandoned agriculture site located in Shahr-e Rey, 5 Km south of Tehran-Iran (Latitude 35° 37' N, Longitude 51° 23' E, 1005 m above sea level). The climate of the site is semi-arid with mild-cold winters and 7 months (Mid April-Mid November) dry season (Fig. 1). Average annual rainfall and average annual temperature are 232 mm and 13.3° C, respectively. The highest rainfall appears in March and the lowest in August. The warmest month occurs in August and the coldest in January. Experiment was conducted at two 4 hectare even-aged (15 years) artificial stand of *Pinus eldarica* Medw. The first stand was irrigated with municipal waste water and the second with well water since plantation. The irrigation was applied daily based on tree water-use and the potential evapo-transpiration, which varied seasonally in response to the climate. The soils of both fields were clay-loam with 32.5% clay, 34.12% silt and 33.38% sand in the field irrigated with municipal waste water and 28.52% clay, 36% silt and 35.48% sand in the field irrigated with well water

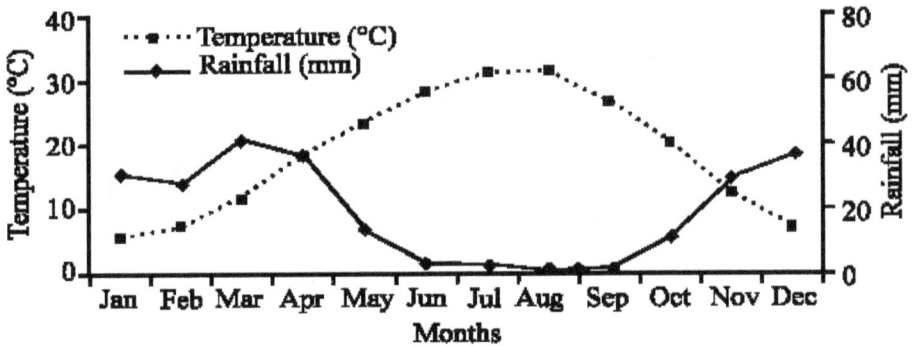

Fig. 1. Embrothermic curve of the study site

The study was established in October 2006. Data was collected using technique of systematic random sampling (Jayaraman, 2000) with 4 replications in either of both fields. Therefore, four plots were identified in each field of irrigated with municipal waste water and well water. Plots were 30 m × 30 m, with tree spacing of 3 m × 4 m (833/ha). In each plot, diameter at breast height (d.b.h.), total height, crown length and crown diameter of total trees were measured and basal area computed. Standing volume of each tree was determined by using form factor (~0.5) and formula $V = 0.4 \times D^2 \times H$, made by Zobeyri

(1994). Where, D is diameter at breast height (d.b.h.), H is total height and V is standing volume.

In each plot, four trees were selected and at the end of growing season needle samples of *P. eldarica* trees taken from the top of crown and the part affected by sunlight (Letacon, 1969; Habibi Kaseb, 1992). This collection provided 16 needle samples in each treatment. At the end of the sampling, one representative needle sample from each plot (by mixing of four samples of each plot) was taken (due to decreasing of samples quantity for chemical analysis). Municipal waste water and well water were sampled daily (3 days in each month) from early June to late November, at three times per day (morning, noon and evening) to make a composite sample of each day.

Water samples were brought to the laboratory in resistant plastic bottles to avoid adherence to the container wall. They were filtered through 42 mm filter paper and stored at 4 °C to minimize microbial decomposition of solids (Yadav *et al.*, 2002; Bhati and Singh, 2003). Several parameters were measured separately, pH and EC by the procedure described using OMA (1990), NH_4-N, NO_3-N, PO_4-P, K, Ca, Mg and Na as per the method given by APHA (1992) and Yadav *et al.* (2002).

Fresh weight of some needles from each treatment was recorded immediately after harvest. Dry weight was recorded after oven drying of needles for 72 h at 80 °C (Bhati and Singh, 2003). Samples of needle were washed using tap water, rinsed with distilled water, oven dried at 80 °C for 24 h (Singh and Bhati, 2005), ground in a stainless steel mill and retained for mineral analysis. For determination of macro and micro-nutrients exception P and N, the needle samples were wet digested as per Jackson (1973) and estimated using an Atomic Absorption Spectrophotometer (AAS). Measurement of P content was performed after a wet digestion using UV–VIS spectrophotometer at 450 nm (Singh and Bhati, 2005). The N content of needle samples digested in concentrate sulfuric acid was determined by the Kjeldahl method (Bhati and Singh, 2003; Bozkurt and Yarilga, 2003).

Average growth parameters and needle nutrients of two irrigation treatments (T_1: irrigation by municipal waste water; T_2: irrigation by well water) were compared using independent-samples t-test. The variations in characteristics of municipal waste water and well water were firstly tested for normality using Shapiro-Wilk's test and then by independent-samples t-test. All the data were analyzed using the SPSS statistical package.

3. Results and discussion

3.1 Waste water and well water

Results indicated that the waters were alkaline in reaction (Table 1). The pH of the municipal waste water in various months ranged from 7.51 to 7.75 and for well water 6.69 to 7.62. Based on results of Patel *et al.* (2004), in our examination the tolerance limit of pH for irrigation ranged from 6.0 to 9.0. The electrical conductivity (EC) of municipal waste water ranged from 1.78 to 2.12 dS m^{-1} with the greatest value detected in August. Average EC of municipal waste water (mean of 18 samples) exceeded 1 dS m^{-1} (1.91 dS m^{-1}) indicating the waste water was saline in nature (Rattan *et al.*, 2005). The pH and EC of the municipal waste water were greater than those of the well water. The concentration of all the nutrient elements was higher in municipal waste water, with NO_3-N content (1.63 mg l^{-1}) being 6.8 times the content in well water (0.24 mg l^{-1}). The content of NH_4-N in municipal waste water (9.05 mg l^{-1}) was also 4.2 times the content in well water (2.15 mg l^{-1}). On average, available content of PO_4-P, K$^+$, Ca^{2+}, Mg^{2+}, Na$^+$ in municipal waste water were greater compared to those in the well water. The most nutrients concentration of municipal waste water were reduced in autumn and increased in summer because of high temperature and evaporation losses of water (Singh and Bhati, 2005).

Although municipal waste water elevated significantly ($P < 0.01$) in all values compared to well water, but the analysis showed that pH, EC, NO_3-N, PO_4-P, K^+, Na^+ of well water samples were within the limits as per the standard prescribed for land disposal and should not pose any serious hazard according to threshold values of WHO (Hach, 2002). However, the contents of NH_4-N and Ca^{2+} of municipal waste water and well water and Mg^{2+} of municipal waste water were on the higher side (Table 1).

Parameters	Municipal waste water		Well water		WHO*
	Range (Min.-Max.)	Mean ± SE	Range (Min.-Max.)	Mean ± SE	
pH	7.51 - 7.75	7.63 ± 0.01 [a]	6.69 - 7.62	7.32 ± 0.05 [b]	6.5 - 8.5
EC (dS m⁻¹)	1.78 - 2.12	1.91 ± 0.02 [a]	0.54 - 0.67	0.590 ± 0.008 [b]	3
NH_4-N (mg l⁻¹)	8.1 - 10.24	9.05 ± 0.11 [a]	1.83 - 2.49	2.15 ± 0.19 [b]	1.5
NO_3-N (mg l⁻¹)	1.58 - 1.89	1.63 ± 0.09 [a]	0.19 - 0.33	0.24 ± 0.08 [b]	3
PO_4-P (mg l⁻¹)	11.45 -14.13	12.69 ± 0.16 [a]	4.62 - 5.64	5.03 ± 0.01 [b]	-------- -
K (mg l⁻¹)	33.06 - 46.31	39.93 ± 0.83 [a]	17.48 - 22.75	19.72 ± 0.36 [b]	-------- -
Ca (mg l⁻¹)	235.54 - 296.20	255.22 ± 4.57 [a]	66.70-101.57	96.77 ± 1.26 [b]	75
Mg (mg l⁻¹)	100.9 - 124	109.85 ± 1.83 [a]	28.9 - 42	35.22 ± 0.79 [b]	50
Na (mg l⁻¹)	135.90 - 150.22	140.45 ± 0.20 [a]	30.18 - 41.03	35.18 ± 0.13 [b]	200

Different superscripts in row indicate significant ($P < 0.01$) difference. Values are mean of eighteen replications (3 days × 6 months) with ± SE; * World Health Organization (WHO): Hach, 2002

Table 1. Characteristics of municipal waste water and well water

3.2 Tree growth

Irrigation with municipal waste water for 15 years produced the largest trees in this treatment. The most frequent trees were found at diameter class of 20 cm and 14 cm, respectively grown on field irrigated with municipal waste water and well water (Fig. 2). In fact, tree growth was greater ($P < 0.01$) in the field irrigated using municipal waste water than in plots irrigated with well water, as indicated by the 17.95 ± 1.33 cm diameter at breast height, 10.04 ± 0.15 m height, 8 ± 0.27 m crown length, 2.53 ± 0.17 m crown average diameter, 264.20 ± 30.02 cm² basal area and 0.139 ± 0.013 m³ standing volume of the trees in waste water irrigated field (Table 2). Similarly, an increase in the growth of olive (*Olea europaea*) trees due to irrigation with municipal waste water has been reported by Aghabarati *et al.* (2008). The study of Stewart *et al.* (1990) also suggested that the addition of municipal waste water on *Eucalyptus grandis* has been resulted in a doubling of growth rate when compared to *E. grandis* grown in a rain fed site in four years.

The increased growth may be linked to sufficient availability of water and better status of nutrients in soil (Larchevêque *et al.*, 2006). Since municipal waste water contains plant nutrients and organic matter, it may improve the properties of soil for increase in growth and biomass production (Guo *et al.*, 2002; Egiarte *et al.*, 2005; Lopez *et al.*, 2006). The increase in growth indicates that waste water application influenced the physiological processes, facilitated early needle initiation and resulted in a net increase in the number of needles. An increase in needles could have captured more solar energy for metabolic use, fixed more CO_2, and produced greater photosynthesis, and growth. This hypothesis is supported by Ceulemans *et al.* (1993) and Myers *et al.* (1996).

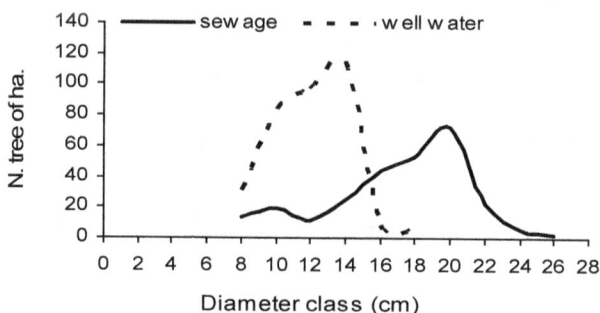

Fig. 2. Distribution of diameter classes for *P. eldarica* trees in two irrigation types

Irrigation type	Diameter at breast height (cm)	Height (m)	Crown length (m)	Crown diameter (m)	Basal area (cm²)	Standing volume (m³)
Trees irrigated with waste water	17.95 [a] (1.33)	10.04 [a] (0.15)	8.0 [a] (0.27)	2.53 [a] (0.17)	264.20 [a] (30.02)	0.139 [a] (0.013)
Trees irrigated with well water	13.50 [b] (0.5)	9.02 [b] (0.10)	7.3 [b] (0.12)	1.90 [b] (0.20)	135.0 [b] (20.5)	0.65 [b] (0.09)

-Different superscripts in column indicate significant difference of each tree attribute between two irrigation types.
-Values in parenthesis are ± SE.

Table 2. Effect of municipal waste water and well water on growth of *P. eldarica* trees

3.3 Mineral composition of needles

The application of municipal waste water significantly increased the macro-elements (N, P, K, Ca, Mg, Na concentration of *P. eldarica* trees needle as compared with well water (Table

3). Increases in minerals concentration may have been due to the effect of nutrients addition through municipal waste water (Meli *et al.*, 2002). This result is in agreement with Singh and Bhati (2005) and Aghabarati *et al.* (2008), whereas a substantially greater above-mentioned minerals concentration were observed in leaf of *Dalbergia sissoo* seedlings and *Olea europaea* trees irrigated with municipal waste water compared to control. However, Guo *et al.* (2002) and Aghabarati *et al.* (2008) had also suggested that a decrease of Mg and Ca, and no difference of Na concentration in leaf of eucalypt and olive tree were treated by municipal waste water. In fact, quantity of nutrients absorption using plant depends upon the total quantity of the nutrients applied through waste water application, soil properties and type of plant (Bozkurt and Yarilga, 2003). The minerals concentration of needle may be ranked from greatest to least as N > Ca > K > Mg > P > Na.

	N	P	K	Ca	Mg	Na
	---------------------------- gr kg $^{-1}$ ----------------------------					
Soil treated with T_1	16. 41[a] (0.27)	0.865 [a] (0.058)	5.79 [a] (0.50)	6.08 [a] (0.27)	1.51[a] (0.12)	0.320 [a] (0.027)
Soil treated with T_2	15.47 [b] (0.35)	0.710 [b] (0.014)	4.49 [b] (0.42)	4.64 [b] (0.26)	1.28 [b] (0.11)	0.198 [b] (0.034)
p-value	<0.01	<0.05	<0.01	<0.01	<0.05	<0.01
Range[*]	5-30	1-5	3-30	10-40	1-7	------

Abbreviations: T_1: municipal waste water; T_2: well water; values are mean of four replications with ± SD in parentheses; different superscripts in column indicates significant difference between T_1 and T_2; [*] Salardini (1992)

Table 3. Mineral composition of *P. eldarica* trees needle by affected by municipal waste water and well water

4. Conclusion

Our study displayed that all growth parameters measured in *P. eldarica* trees were statistically greater in effluent-irrigated area than in well-watered area. As a whole, the use of municipal effluent in irrigations can be an overflowing resource from the nutrient elements required for plants (Yadav *et al.*, 2002; Mapanda *et al.*, 2005; Toze, 2006). As a matter of fact, high nutrient concentrations in effluent, compared to those in well water, cause the nutrient accumulation in the soil (Stewart and Flinn, 1984; Phillips *et al.*, 1986; Stewart *et al.*, 1990; Keller *et al.*, 2002; Selivanovskaya *et al.*, 2002; Emongor and Ramolemana, 2004) and makes easy the access of plants to the high nutrient concentration (macro and micro elements) and increases their growth. Accordingly, in agreement with our findings the results of Stewart and Flinn (1984, on *Pinus eldarica*), Phillips *et al.* (1986, *Pinus eldarica*), Ostos *et al.* (2007, on *Pistacia lentiscus*) show that faster growth of tree occurs in the effluent-irrigated areas. This is mostly due to high nutrient concentration in effluent. It may be also noted that the nutrient contents in the municipal effluent is more than needed by plants whereas in the such conditions trees can produce greater biomass (Fitzpatrick *et al.*, 1986; Martinez *et al.*, 2003; Sing and Bhati, 2005; Guo *et al.*, 2006). Regarding the differences indicated above and positive effects of effluent on the growth of *P. eldarica*, it can be recommended that the produced huge municipal effluent in south of Tehran can be used for accomplishment of plantation projects and for development of rural and urban green spaces and green belts around the city and for reduction of air pollution, too. It is necessary to clarify that the decision for each location should be made based on accurate management, chemical, physical and microbial characteristics of water, soil and plant, according to international standards.

5. Acknowledgement

Authors are thankful to Natural Resources Faculty of Tarbiat Modares University for providing research facilities and funding of this research and to Department of Forestry for technical and scientific assistance. We gratefully acknowledge Shahr-e-Ray Municipality for their support on field assistance of this research.

6. References

Aghabarati, A.; Hosseni, S.M, Esmaeili, A. & Maralian, H. (2008). Growth and mineral accumulation in *Olea europaea* L. trees irrigated with municipal effluent. *Res. J. Environ. Sci.*, 2 (4): 281-290.

Al-Jamal, A.M.S.; Sammis, T.W.; Mexal, J.G.; Picchioni G.A. & Zachritz, W.H. (2000). A growth irrigation scheduling model for wastewater use in forest production. *Agricultural Water Management*, 56: 57-79.

APHA. (1992). *Standard Methods For The Examination OF Water And Wastewater*. APHA, AWWA and WPCF. 16th ed.

Bhati, M. & Singh, G. (2003). Growth and mineral accumulation in *Eucalyptus camaldulensis* seedlings irrigated with mixed industrial effluents. *Bioresource Technol.*, 88: 221-228.

Bozkurt, M.A. & Yarilga, T. (2003). The effects of waste water sludge applications on the yield, growth, nutrition and heavy metal accumulation in apple trees growing in dry conditions. *Turk. J. Agric. For,.* 27: 285-292.

Brar, M.S., S.S. Mahli, A.P. Singh, C.L. Arora and K.S. Gill, 2000. Sewer water irrigation effects on some potentially toxic trace elements in soil and potato plants in northwestern India. *Can. J. Soil Sci.*, 80: 465-471.

Ceulemans, R.J.; Pontailler, F.M. & Guittet, J. (1993). Leaf allometry in young poplar stands: reliability of leaf area index estimation, site and clone effects. *Biomass Bioenerg.*, 4: 769-776.

Cromer, R.N., P. Tompkins and N.J. Barr, 1987. Irrigation of *Pinus radiata* with waste water: tree growth in response to treatment. *Aus. Forest Res.*, 13: 57-65.

Datta, S.P.; Biswas, D.R.; Saharan, N.; Ghosh, S.K. & Rattan, R.K. (2000). Effect of long-term application of sewage effluents on organic carbon, bioavailable phosphorus, potassium and heavy metals status of soils and uptake of heavy metals by crops. *J. Indian Soc. Soil Sci.*, 48: 836-839.

Egiarte, G.; Arbestian, C.M.; M., Alonso, A.; Rui´Z-Romera E. & Pinto, M. (2005). Effect of repeated applications of waste water sludge on the fate of N in soils under Monterey pine stands. *For. Ecol. Manage.* 216: 257-269.

Emongor, V.E. & Ramolemana, G.M. (2004). Treated sewage effluent (water) potential to be used for horticultural production in Botswana. *Physics and Chemistry of the Earth*, 29: 1101–1108.

Fitzpatrick, G.E., Donselman H. & Carter, N.S. (1986). Interactive effects of sewage effluent irrigation and supplemental fertilization on container - grown trees. *Hort. Science*, 21(1): 92–93.

Guo, L.B.; Sims, R.E.H.; Horne, D.J. (2002). Biomass production and nutrient cycling in *Eucalyptus* short rotation energy forests in New Zealand. I: biomass and nutrient accumulation. *Bioresource Technol.*, 85: 273-283.

Guo, L.B.; Sims, R.E.H. and Horne, D.J. (2006). Biomass production and nutrient cycling in Eucalyptus short rotation energy forests in New Zealand: II. Litter fall and nutrient return. *Biomass and Bioenergy*, 30: 393–404.

Gupta, A.P.; Narwal, R.P.; Amtil, R.S. (1998). Sewer water composition and its effect on soil properties. *Bioresource Technol.*, 65: 171-173.

Habibi Kaseb, H. (1992). *Forest Pedology*. Tehran University Press, 424 pp.

Hach, C. (2002). *Water Analysis Handbook*, Loveland, Colorado, USA, p. 61-62.

Jackson, M.L. (1973). *Soil Chemical Analysis*. Prentice Hall of India Private Ltd., New Delhi.

Jayaraman, K. (2000). *A Statistical Manual for Forestry Research*. FORESPA Publication., 240 pp.

Karpiscak, M.M.; Gerba, C.P.; Watt, P.M.; Foster K.E. & Falabi, J.A. (1996). Multi-species plant systems for wastewater quality improvement and habitat enhancement. International association on water quality. *Water Sci. Technol.*, 33: 231–236.

Keller, C.; Grath, S.P.Mc. & Dunham, S.J. (2002). Trace metal leaching through a soil grassland system after sewage sludge application. *J. Environ. Qual.*, 31: 1550-1560.

Larchevarcheveque, M.; Ballini, C., Korboulewsky N. & Montes, N. (2006). The use of compost in afforestation of Mediterranean areas: Effects on soil properties and young tree seedlings. *Sci. Total Environ.*, 369: 220-230.

Letacon, F. (1969). Une methode originale de prelevemennts foliaires, *R.F.F.* 3: 196-197.

Lopez, A.; Pollice, A.; Lonigro, A.; Masi, S.; Palese, A.M.; Cirelli, G.L.; Toscano, A. & Passino, R. (2006). Agricultural wastewater reuse in southern Italy. *Desalination*. 187: 323-334.

Madejon, P.; Maranon, T. & Murillo, J.M. (2006). Biomonitoring of trace elements in the leaves and fruits of wild olive and holm oak trees. *Sci. Total Environ.* 355: 187-203.

Martinez, F.; Cuevas, G.; Calvo R. & Walter, I. (2003). Biowaste effects on soil and native plants in semiarid ecosystem. *J. Environ. Qual.*, 32: 472–9.

Meli, S.; Porto, M.; Bellingo, A.; Bufo, S.A.; Mazzatura, A. & Scopa, A. (2002). Influence of irrigation with lagooned urban wastewater on chemical and microbiological soil parameters in a citrus orchard under Mediterranean condition. *Sci. Total Environ.*, 285: 69-77.

Mosadegh, A. (1999). *Plantation and Forest Nurseries*. University of Tehran Press. 516 pp.

Myers, B.J.; Theiveyanath, S.O.; Brian, N.O. & Bond, W.J. (1996). Growth and water use of *Eucalyptus grandis* and *Pinus radiata* plantations irrigated with effluent. *Tree Physiol.*, 16: 211-219.

Naghshinehpour, B. (1998). Application of effluent in agriculture productions and soil rehabilitation. First congress on the programming and policy in infrastructural matter (water and soil), Ministry of Agriculture.

Neilson, G.H.; Stevenson, D.S.; Fitzpatrick, J.J. & Brownlee, C.H. (1989). Nutrition and yield of young apple trees irrigated with municipal waster water. *J. Am. Soc Hortic. Sci.*, 114: 377-383.

OMA, (1990). *Official Methods of Analysis*. 15th ed. Association of Official Analytical Chemists, Arlington, Virginia, USA.

Ostos, J.C.; Pez-Garrido, R.Lo.; Murillo, J.M. & Lo´pez, R. (2007). Substitution of peat for municipal solid waste- and sewage sludge-based composts in nursery growing media: Effects on growth and nutrition of the native shrub *Pistacia lentiscus* L. *Bioresource Technology*.

Patel, K.P.; Pandaya, R.R.; Maliwal, G.L.; Patel, K.C.; Ramani, V.P. & George, V. (2004). Heavy metal content of different effluents and their relative availability in soils irrigated with effluent waters around major industrial cities of Gujarat. *J. Ind. Soc. Soil Sci.*, 52: 89-94.

Phillips, R.; Fisher J.T. & Mexal, J.G. (1986). Fuelwood production utilizing *Pinus eldarica* and sewage sludge fertilizer. *Forest Ecology and Management*, 16: 95-102.

Ramirez-Fuentese, Lucho-Constsntino, C.; Escamilla-Silva, E. & Dendooven, L. (2002). Characteristics and carbon and nitrogen dynamics in soil irrigated with waste water for different lengths of time. *Bioresource Technol.*, 85: 179-187, 2002.

Rattan, R.K.; Datta, S.P.; Chhonkar, P.K.; Suribabu, K. & Singh, A.K. (2005). Long-term impact of irrigation with waste water effluents on heavy metal content in soils, crops and groundwater-a case study. *Agr. Ecosyst. Environ.* 109: 310-322.

Salardini, A. (1992). *Soil Fertility*. Tehran University Press, 440 pp.

Selivanovskaya, S.Y.; Latypova, V.Z.; Kiyamova S.N. & Alimova, F.K. (2001). Use of microbial parameters to access treatment methods of municipal sewage sludge applied to grey forest soils of Tatars tan. *Agriculture, Ecosystem and Environment*, 86: 145-153.

Sharma, A. & Ashwath, N. (2006). Land disposal of municipal effluents: Importance of choosing agroforestry systems. *Desalination*, 187: 361-374.

Sharma, R.K.; Agrawal, M. & Marshall, F. (2007). Heavy metal contamination of soil and vegetables in suburban areas of Varanasi, India. *Ecotox. Environ. Safe.*, 66: 258-266.

Singh, G. & Bhati, M. (2005). Growth of *Dalbergia sissoo* in desert regions of western India using municipal effluent and plant chemistry. *Bioresource Technology*, 69: 1019-1028.

Stewart, H.T.L. & Flinn, D.W. (1984). Establishment and early growth of trees irrigated with wastewater at four sites in Victoria, Australia. *Forest Ecology and Management*, 8: 243-256.

Stewart, H.T.L.; Hopmans, P.; Flinn, D.W. and Hillman, T.J. (1990). Nutrient accumulation in trees and soil following irrigation with municipal effluent in Australia. *Environmental Pollution*, 63: 155-177.

Tabatabaei, M. (1998). Sustainability in irrigation designs with wastewater. *Journal of water and Environment*, 31: 28-35.

Tajrishi, M. (1998). New and comprehensive outlook to the problem of municipal effluent of Tehran. *The Journal of Water and Effluent*, 28: 16-30.

Torabian, A. & Hashemi, F. (1999). Irrigation of green space with treated wastewater of Tehran. *The Journal of Water and Effluent*, 29: 31-36.

Toze, S. (2006). Reuse of effluent water-benefits and risks. *Agr. Water Manage.*, 80: 147-159.

Wheeler, P.M.; Edmeades, D.C. & Christie, R.A. (1992). Effect of aluminum on relative yield and plant chemical concentrations of cereals grown in solution culture at low ionic strength. *J. Plant Nutr.*, 15: 403-418.

Yadav, R.K.; Goyal, B.; Sharma, R.K., Dubey, S.K. & Minhas, P.S. (2002). Post-irrigation impact of domestic waste water effluent on composition of soils, crops and ground water-a case study. *Environ. Int.*, 28: 481-486.

Zobeyri, M. (1994). *Forest Inventory.* Tehran University Press, 401 pp.

4

Effects of Reclaimed Water on Citrus Growth and Productivity

Dr. Kelly T. Morgan
University of Florida, Soil and Water Science Department,
Southwest Florida Research and Education Center,
Immokalee, FL 34142
USA

1. Introduction

Sewage wastewater or effluent is often viewed as a disposal problem. However, it can be a source of water for irrigation, creating an alternative disposal method for wastewater treatment facilities, benefiting agriculture as an alternate source of irrigation water, and reducing the demand for use of surface or ground water for irrigation (Parsons et al., 2001a and b). Treated wastewater, also known as reclaimed water, is typically treated municipal sewage from which excess plant nutrients, organic compounds and pathogens have been removed. The terms wastewater, treated wastewater and reclaimed water will be used interchangeably in this chapter.

The characteristics and treatment of these treated waters will be described and discussed in this chapter along with use as an irrigation source for citrus production. Potential disadvantages of using reclaimed water for agricultural irrigation include real or perceived concerns about reductions in surface and ground water quality (i.e. nutrients and heavy metals), harmful effects on workers that come in contact with treated wastewater (i.e. organic compounds and pathogens), and the safety of crops for human consumption (i.e. carcinogens and pathogens) (Parsons & Wheaton, 1996; Parsons et al., 1995). In some arid regions where freshwater supplies are limited, irrigation with reclaimed water is already commonly practices (Feigin et al., 1991). Israel was a pioneer in the development of wastewater re-use practices, but was quickly followed by many other countries (Angelakis et al., 1999). Israel and the Palestinian Autonomous Regions, for example, are projected to use 3500 million m³ of water in 2010, with 1400 million m³ (40% total water supply) used for irrigation. Treated sewage water used for irrigation would be approximately 1000 million m³ or 70% of agricultural water demand and will play a dominant role in sustaining agricultural development (Haruvy, 1994). Wastewater is a preferred marginal water source, since its supply is reliable and uniform, and is increasing in volume due to population growth and increased awareness of environmental quality (Haruvy & Sadan, 1994). Costs of this water source are low compared with those of other unconventional irrigation water sources (e.g. desalinization) since agricultural reuse of urban wastewater serves also to dispose of treated urban sewage water (Haruvy & Sadan, 1994). Total cost of supplying wastewater for agricultural reuse (i.e. treatment, storage and conveyance costs) minus total costs of alternative safe disposal (e.g. deep well injection and wetlands creation) must be considered when developing wastewater reuse systems (Angelakis et al., 1999; Arora&Volutchkov, 1994; Haruvy, 1997)).

2. Wastewater reuse: the general case

The rapid development of irrigation has resulted in an increased water demand. Accessible water resources (e.g. rivers and shallow aquifers) in most agricultural areas are now almost entirely committed (Angelakis et al., 1999). Alternative water resources are therefore needed to satisfy further increases in demand. This is particularly a necessity in regions which are characterized by severe mismatches between water supply and demand. Low water resource availability and temporal symmetries in availability result in water provided for human consumption and other urban use with less water for agricultural use. The reduction in water availability for agriculture can lead to reduced sustainability of agricultural enterprises and/or environmental problems (Angelakis et al., 1999). One potential alternate source of irrigation water for agriculture situated near large urban centers is treated wastewater. Reclaimed water contains many nutrients essential for plant growth, and may have an effect similar to that of frequent fertigation with a dilute concentration of plant nutrients (Neilsen et al., 1989). In addition, recycling these nutrients may prevent pollution of surface or ground water (Sanderson, 1986).

In the Mediterranean basin, wastewater has been used as a source of irrigation water for centuries. In addition to providing a low cost water source, the use of treated wastewater for irrigation in agriculture combines three advantages 1) using the fertilizing properties of the water can partially eliminate synthetic fertilizers demand and contribute to decrease nutrient concentration of rivers, 2) the practice increases the available agricultural water resources, and 3) in some areas, it may eliminate the need for expensive tertiary treatment (Angelakis et al., 1999).

In a review by Haruvy (1997) wastewater quality or treatment levels are defined by various constituents such as 1) organic matter- biochemical oxygen demand, chemical oxygen demand and total suspended solids; 2) organic pollutants (i.e. stable organic matter that may affect health); 3) trace elements resulting from industrial water use; 4) pathogenic microorganisms; 5) potential plant nutrients (e.g. N and P); and 6) salinity. Treatment processes are generally divided into primary, secondary and advanced or tertiary processes. Primary treatment includes basic treatment such as screening to remove coarse solids and solid precipitates. Secondary treatment includes low-rate processes (e.g. stabilization or sediment ponds) with high land and low capital and energy inputs, and high rate processes (e.g. activated sludge) with low land and high capital and energy inputs (Pettygrove & Asano, 1985). Tertiary stages of treatment further improve water quality by nitrification and denitrification to reduce water N.

Environmental hazards may be caused by each constituents (e.g. nutrients, heavy metals) left in wastewater and may leach below the root zone increasing groundwater pollution (Feigin et.al, 1990). Salinity of reclaimed water is generally within acceptable ranges and often lower than other irrigation sources, however, salinity levels may be acceptable only for ground application and not direct plant contact in some treatments processes (Basiouny, 1982). Leaching of fertilizers, pesticides and salts from soils irrigated with treated wastewater or over application of poor quality wastewater has resulted in the progressive loss of subsurface water quality and decrease in groundwater resources in some areas (Lapena et al., 1995). However, when properly managed, the use of treated wastewater in agriculture to conserve water resources and to safely and economically dispose of wastewater is a very feasible option.

Treated municipal wastewater has become an important potential source of irrigation and plant nutrients and has been used successfully in the production of high yield marketable quality crops for decades (Allen & McWhorter, 1970; Crites, 1975; Day, 1958; Henry et al., 1954; Stokes et al., 1930). The response of plants and soils to municipal treated effluent is dependent on the quality of the applied effluent and nature and efficiency of the wastewater treatment, with generally higher treated water resulting in the best growth and yields (Basiouny, 1984). Recently, wastewater has been used to increase yield and improve quality of grain crops (Al-Jaloud et al., 1993; Day & Tucker, 1977; Day et al., 1975; Karlen et al., 1976; Morvedt & Giovdane, 1975; Nguy, 1974), cotton (Bielorai et al., 1984; Feigin et al., 1984), forage (Bole & Bell, 1978) and vegetable crops (Basiouny, 1984; Kirkham, 1986; Neilsen et al., 1989 a, b, c, 1991; Ramos et al., 1989). Reclaimed water has been successfully used to irrigate many fruit crops; apples (Nielsen et al., 1989a), cherries (Nielsen et al., 1991), grapes (Neilsen et al., 1989a), peaches (Basiouny, 1984) and citrus (Esteller et al., 1994; Kale & Bal, 1987; Koo & Zekri, 1989; Morgan et al., 2008; Omran et al., 1988; Wheaton & Parsons, 1993; Zekri & Koo, 1990).

3. Guidelines for wastewater reuse in irrigation

The Ganga is the most important river system in India. It rises from the Gangotri glacier in the Himalaya mountains at an elevation of 7138 m above mean sea level as a pristine river. Half a billion people (almost one tenth of the world's population) live within the Ganga river basin at a average density of over 500 per km^2 (Singh et al., 2003). This population is projected to increase to over one billion people by 2030. Sewage treatment plants provide agricultural benefits by supplying irrigation and nonconventional fertilizers in the Ganga river basin as an alternate disposal of effluent into the river (Singh et al., 2003). Areas with extensive agriculture and rapidly escalating population must use water resources in a sustainable way and require guidelines to insure the health of the population and maintain water quality and the environment in sensitive areas such as the Gana river basin.

Wastewater reuse guidelines typically cover four areas for each application (i.e. type of crop irrigated): chemical standards, microbiological standards, wastewater treatment processes and irrigation techniques (Angelakis et al., 1999). The degree of treatment required and the extent of monitoring necessary depend on the specific use and crop. In general, irrigation systems are categorized according to the potential degree of human exposure.

The highest degree of treatment is always required for irrigation of crops that are consumed uncooked (Angelakis et al., 1999). However, wastewater is often associated with environmental and health risks. As a consequence, its acceptability to replace other water resources for irrigation is highly dependent on whether the health risks and environmental impact entailed are acceptable. Evaluation of reusing wastewater is the quality of the water in terms of the presence of potentially toxic substances or of the accumulation of pollutants in soil and crops. It is important to access the source of the wastewater for heavy metals from industries or synthetic chemicals normally present in urban wastewater (e.g. oils, disinfectants). There have also been debates about applicable microbiological practices and the type of crops that should be irrigated with treated effluent (Asano & Levine, 1996). One set of guidelines established in California, USA and now accepted nationwide and other countries of the world, promote very high water quality standards (comparable to drinking water standards), confident that costly treatment practices provide safe enough water (i.e. free of enteric viruses and parasites) for who can afford it. The "California criteria" (State of California, 1978) stipulate conventional biological wastewater treatment followed by tertiary treatment, filtration and chlorine disinfection to produce effluent that is suitable for irrigation (Arora & Voutchkov, 1994). In support of this approach, Asano & Levine (1996) reported two major epidemiological studies conducted in California during the 1970s and 1980s. These studies scientifically demonstrated that food crops irrigated with municipal wastewater reclaimed according to the California approach could be consumed uncooked without adverse health effects. However, the nutrients removed by the tertiary treatment are not available for fertilizing of the crops.

In contrast to the California approach, the guidelines produced by the World Health Organization (WHO) are less stringent and require a lower level of water treatment (WHO, 1989). The WHO guidelines are, however, more restrictive in assuring microbiological quality of treated water, requiring monitoring of fecal Coliform bacteria (also required in the California criteria) as well as human parasitic nematodes.

Outside of Europe, other countries (e.g. Israel, South Africa, Japan and Australia) have chosen criteria more or less similar to those adopted by California (and elsewhere in the US). Most countries in Europe accept the 1989 HWO guidelines but contain additional criteria such as treatment requirements and/or use limitations in order to ensure proper public health protection. The California approach has the most data in its own support and thus has been accepted by more countries because of its "safety first" philosophy but is the most expensive to implement.

4. Risk assessment

Shuval et al. (1997) developed a preliminary model for the assessment of risk of infection and disease associated with wastewater irrigation of vegetables eaten uncooked based on a modification of the Haas et al. (1993) risk assessment model for drinking water. The modifications included determining the amount of wastewater that would cling to irrigated vegetables and estimates of the concentration of pathogens that would be ingested by consuming vegetables irrigated with wastewater of different propagule densities. The model was validation with data from a cholera epidemic caused in part by consumption of wastewater irrigated vegetables and provided reasonable approximation of the levels of disease that really occurred. The risk assessment, using this model, of irrigation with treated wastewater effluent meeting the WHO guidelines (WHO, 1989, 1,000 fecal coliform bacteria 100 ml^{-1}) indicates the risk of contracting a virus disease is about 10^{-6} to 10^{-7}. Regli et al. (1995) concluded that guidelines for drinking water standards should be designed to ensure that human populations are not subjected to the risk of infection by enteric disease at $> 10^{-4}$ for a yearly exposure. Thus this preliminary study suggested that the WHO guidelines provided a factor of safety some 1 to 2 orders of magnitude greater than that called for by the United States Enviornmental Protection Agency (USEPA) (USEPA, 1992) for microbial standards for drinking water.

5. Wastewater irrigation of Florida citrus: a case study

Florida has experienced rapid growth in population during the last 50 years with a 5.5-fold population increase from 1950 to 2000 (U.S. Census Bureau, 1997; Perry & Mackum, 2001;

Smith, 2005). Groundwater withdrawal for domestic and irrigation use has increased by 15.5 and 20.7 times, respectively, during the same period (Marella & Berndt, 2005). Likewise, the amount of wastewater generated by cities in Florida has increased more than five-fold since 1950. Environmental concerns about degradation of surface waters by treated effluent water have caused many communities to consider advanced secondary treated wastewater (reclaimed water) reuse. Currently there are 440 reclaimed water reuse systems in Florida irrigating 92,345 ha with 2,385 million liters of reclaimed water per day (Florida Department of Environmental Protection, 2005). The majority of these systems irrigate golf courses, public right-of-ways, and home landscapes. However, 6,144 ha of production agriculture are currently irrigated with reclaimed water, with citrus (*Citrus* spp L.) orchards accounting for all but 364 ha (Morgan et al., 2008).

Florida citrus production benefits from irrigation because the average annual rainfall of more than 1200 mm is unevenly distributed throughout the year with approximately 75% of annual rainfall occurring from June to September (Koo, 1963). Furthermore, Florida citrus trees are grown on sandy soils with very low water holding capacity, particularly orchards in the central "ridge" portion of the state. Typical available water content values for central Florida ridge citrus soils range from 0.05 to 0.08 cm^3 cm^{-3} (Obreza & Collins, 2003). Increased water use by the growing population and localized water shortages during low rainfall years have resulted in the development of water use restrictions, and decreases in permitted water use for agriculture. Increased use of reclaimed water for agricultural irrigation would not only reduce the wastewater disposal problem for urban areas, but could also reduce the amount of water withdrawn from surficial and Floridan aquifers for irrigation.

Water for irrigation is no longer abundant and restrictions on the use of available groundwater in agriculture are becoming more severe. Due to the increasing demand for water, water use for agricultural purposes has become strictly regulated in Florida (Koo & Zekri, 1989; Wheaton & Parsons, 1993; Zekri & Koo, 1990). Additionally, urban growth, especially in the coastal areas of Florida, has increased the need for efficient and environmentally safe disposal of reclaimed water. The Department of Environmental Regulation (FDER) has restricted the disposal of municipal reclaimed water into lakes, rivers and streams, so alternative disposal sites need to be found (Maurer & Davies, 1993).

Wastewater has been recognized as a possible important source of major plant nutrients (e.g. N, P and K), although the chemical composition of wastewater varies between locations (Berry et al., 1980). Long term studies using reclaimed water to irrigate citrus for up to 60 years in Egypt found no adverse effects on tree growth compared to ground water irrigated citrus (Omran et al., 1988). Similarly, irrigation with reclaimed water increased growth and yield of citrus on well drained sandy soils of the Florida Ridge with no adverse affects on health and yield of mature trees (Koo & Zekri, 1989; Zekri & Koo. 1990). Similar results were observed for young citrus trees grown of well drained soils (Wheaton & Parsons, 1993).

Soil types and drainage patterns of the poorly drained flatwoods soils near the Florida coastline vary considerably due to the presence of a high water table (Maurer & Davies, 1993). The potential waterlogging of the fatwoods hold problems not associated with citrus grown on the Ridge. In a three year study, trees grown of poorly drained sandy soils were irrigated with a simulated reclaimed water, simulated reclaimed water with fertilizer added or ground water with fertilizer added for a period of three years after planting (Maurer & Davies, 1993). Trees irrigated with simulated reclaimed water and ground water with fertilizer added had significantly larger canopies and trunk diameters than trees irrigated with simulated reclaimed water only indicating that use of reclaimed water alone was insufficient to support adequate growth of young citrus trees.

Prior to 1987, the City of Orlando and Orange County wastewater treatment plants discharged their effluent into Shingle Creek, a tributary of Lake Tohopekaliga (Zekri & Koo, 1989). Faced with the need to expand wastewater treatment volume and a state requirement to eliminate discharge of treated effluent to surface waters, the city and county entered a negotiated settlement with the Florida Department of Environmental Regulation and the United States Environmental Protection Agency to cease effluent discharge into Shingle Creek and develop an innovative reclamation program (Zekri & Koo, 1989). Initial funding of $180,000,000 established the project which is called Water Conserv II (Parsons et al., 2001a). The Water Conserv II/Southwest Orange County Water Reclamation Project (Conserv II) involves the use of highly treated wastewater for citrus irrigation and groundwater recharge. It is one of the largest water reuse projects in the United States and the first reuse program permitted in Florida that involves irrigation of crops intended for human consumption. The program, which became fully operational in January, 1987, currently delivers approximately 133,000 cubic meters of reclaimed water per day (cmd) (275,000 cmd maximum flow) to approximately 1750 ha of citrus. Other users of reclaimed water from the Water Conserv II project are eight foliage greenhouse operations, four tree farms, two ferneries, and three golf courses. The reclaimed water is distributed though 80 km of pipelines maintained by the project. Excess reclaimed water is disposed of in 71 ha of rapid infiltration basins that recharge surficial and Floridan aquifers. Water Conserv II is the largest reclaimed water agricultural irrigation project of its type in the world and was the first project in Florida to be permitted to irrigate crops for human consumption with reclaimed water (McMahon et al., 1989).

Citrus groves in western Orange and eastern Lake Counties, Florida (lat. 28o 28' 20" N, long. 81o 38' 50" W, elevation 64 m) were selected for the Conserv II project because of their high demand for irrigation water and soil series which have high permeability. The predominant soil order in this area is Entisol, with Candler fine sand (hyperthermic, uncoated, Typic Quartzipsamment) being the dominant soil series (Obreza & Collins, 2003). The Candler series consists of excessively drained, very rapidly permeable soils formed from marine deposits. These soils are located in upland areas and typically have slopes of 0-12%. The A and E horizons consist of single-grained fine sand, have a loose texture, and are strongly acidic (pH = 4.0 - 5.5). A Bt horizon is located at a soil depth of 2 m and includes loamy lamellae of 0.1 to 3.5 cm thick and 5 to 15 cm long. This area is a primary aquifer recharge area for the lower Florida peninsula (Zekri & Koo, 1989). Use of reclaimed water for irrigation, in lieu of previous surface water discharges, benefited the urban sector by 1) reducing competition from the agricultural demand for potable water and 2) increasing available groundwater supplies through supplementing natural recharge of the aquifer. The agricultural sector benefited from the project by 1) providing citrus growers with a long-term source of water that will increase and not decrease with urban growth and 2) reduced irrigation pumping costs associated with deep wells previously used for irrigation.

To receive reclaimed water for irrigation at no cost, citrus growers were required to sign a contract with the City of Orlando and Orange County to accept 1270 mm of water per year for a period of at least 20 years. Initially, there was grower resistance because of concerns that use of the reclaimed water might damage citrus trees, or make the fruit unmarketable. As part of the contract, the growers requested long-term studies on the effects of reclaimed water on citrus tree health and fruit quality. Orchards were not now required to accept the full 1270 mm of water per year under the contract because rapid infiltration basins (RIBs) were installed in the early 1990s. Due to the highly porous nature of the soils, the RIBs function as alternate disposal sites (particularly during the normally wet summer rainy season) where the reclaimed water is applied at high rates and allowed to percolate to the ground water. Still questions persisted regarding the effect of long-term use of wastewater on tree productivity.

Conserv II water is good quality water having low mineral concentrations and very low TDS (Zekri & Koo, 1990). Characteristics and chemical composition of reclaimed water provided by the Water Conserv II project are summarized in Table 1. This treated wastewater is highly treated having relatively low biological oxygen demand and mineral contents. In general, growers in the project have followed sound irrigation practices (Koo & Zekri, 1989; Zekri & Koo, 1990). An initial survey of orchards receiving reclaimed water from Conserv II was conducted from 1986 to 1989. No adverse affects of reclaimed water use on tree health and productivity were noted in the initial phase of the orchard survey, however, continued monitoring was suggested to determine long term effects (i.e. metal accumulation in soil, leaves or fruit).

Leaf samples indicated that both trees irrigated with reclaimed wastewater and ground water were adequately fertilized. No consistent trends were observed for leaf K, Ca, Mg and Cu contents. Although leaf Na content from trees irrigated with reclaimed wastewater was twice as high as trees on well water, Na content of both groups was well within the optimum standard values for citrus (Obreza & Morgan, 2008). While the surface six inches of soil did not show any consistent trends due to irrigation with reclaimed water, accumulation of nutrient elements became more apparent when the soil profile to one meter was examined. Higher N and P were found in the soil profile of reclaimed water irrigated groves in 1988 when compared to the well water groves. No differences were observed in the extractable soil K, Ca, Mg and Na of reclaimed water and control groves. Fruit from trees irrigated with reclaimed water had lower soluble solids and acid content in 1987 than fruit from control trees. Such effects of irrigation on juice quality are well documented (Koo & McCornack, 1965; Koo & Smajstral, 1984). In 1987, soil water content was considerably higher in the reclaimed water groves than the control groves resulting in lower soluble solids. In 1988, soil water content in the reclaimed groves was only slightly higher than the control groves and differences in soluble solids were not detected.

A long-term replicated small plot study was conducted from 1989 to 2000 to determine the affect of irrigation with reclaimed water on citrus trees on sandy soils and irrigated with water supplied by the Water Conserv II project (Parsons et al., 2001b). Reclaimed water was applied to citrus trees from planting to 10 years of age at 400, 1500 and 2500 mm per year at equal monthly amounts. Ground water applied at recommended rates based on daily evapotranspiration was provided as a control. The highest two treated wastewater irrigation rates promoted greater trunk and canopy growth. In the first three years, trunk diameters were similar for the ground water control and 400 mm rate of reclaimed water. From years four to 10, trees that received the 1250 and 2500 mm per year application rates were significantly larger than those receiving the 400 mm treatment. The 2500 mm per year reclaimed water rate produced well, even though the high irrigation rate caused a significant reduction in juice soluble solids, 19.3% more fruit per hectare than the 400 mm per year treatment resulting in 15.5% total soluble solids per hectare compared with the 400 mm rate because of the greater fruit production at the higher irrigation rate. These results show that irrigation with excessively high rate of reclaimed water was not detrimental to canopy growth and fruit production. This was due to the good drainage of this sandy soil and the lack of root diseases. The slight reduction in juice soluble solids at the high irrigation rate was more than compensated for by the higher total soluble solids yield.

In the same study, leaf N contents were slightly lower in plants irrigated with groundwater than wastewater (Parsons et al., 2001b). It was concluded that this was due to elevated levels of organic matter found in wastewater which provided additional N. Higher leaf N was also found in treated wastewater irrigated sweet-cherry (Neilsen et al.,1991), apples (Neilsen et al., 1989c), cotton (Feigin et al., 1984) and peach trees (Basiouny, 1984). No significant differences in leaf P contents were found between plants irrigated with either groundwater or wastewater, in spite of wastewater supplying a higher soil P load. This is explainable considering that the amount of P supplied by both kinds of irrigation water was a small percentage of total P from soil and fertilizer sources. Leaf K, concentration in leaves of plants irrigated with groundwater was significantly higher than in plants irrigated with wastewater probably because the elevated Na levels in the wastewater inhibited K uptake by citrus plants (Banuls et al., 1990). Soil solution Na has been found to antagonize K uptake in other plants (Epstein, 1961; LaHaye & Epstein, 1969; Cramer et al., 1987). Plants irrigated with wastewater showed higher leaf content of Cl and Na than those irrigated with groundwater. Citrus is considered to be a salt sensitive crop (Mass & Hoffman, 1977) and salinity causes reduction in growth, ionic imbalance, and adverse water relations in citrus (Walker et al., 1982). Embleton et al. (1973) established 0.7% and 0.25% as the limit for Cl and Na concentrations, respectively. Above these tissue concentration limits, toxic effects are manifested in citrus. No salinity effects were observed over the 10 year study because the nearly 950 mm rainfall during Florida's rainy season (June to Septhermber) does not allow for accumulation of salts.

A second orchard monitoring project to determine any adverse effects on citrus tree health and production associated with long-term irrigation using reclaimed water started in 1995 and was terminated in 2004 (Morgan et al., 2008). The objective of this project was to determine whether long-term irrigation with treated municipal wastewater 1) reduced tree health (i.e. canopy appearance and leaf nutrient content), 2) decreased visual fruit loads, 3) impacted internal fruit quality (i.e. Brix, titratable acid, Brix:acid ratio, and/or 4) increased in soil contaminant concentrations. In 1994, 10 orchards irrigated with one of the two water sources were selected for a total of 20 orchards. These 20 orchards were paired so that trees of the same scion and relative age were irrigated with either water sources. The scions used were 'Hamlin' and 'Valencia' oranges (C sinensis L.), 'Sunburst' tangerine (C. reticulata Blanco), and 'Orlando' tangelo (C. reticulata Blanco x C. paradisi Macfadyn) however, the root stocks were not always consistent among the two water sources. Random trees over a four hectare plot in each orchard were evaluated quarterly for canopy appearance, leaf color, fruit crop, and weed cover. Each orchard received a separate visual rating for each category on a 1-5 scale. A rating of 1 indicates a less dense canopy compared with visual inspection of orchards in the area at the same time period, leaf color would be chlorotic and/or have visual deficiency symptoms, the fruit crop would be low enough to be unharvestable, and the weed population would be very low indicating insufficient nutrition, soil water content or excess herbicide application. Ratings of 5 would indicate a thick dense canopy with excessive vegetative growth, dark green leaves with N concentrations above that considered optimum, a fruit crop considered to be well above the average for trees of comparable age and size in the area, and a dense weed population in the herbicide zone well in excess of standard grower practices.. Fruit samples (20 fruit) were taken from five trees in each orchard just prior to harvest and analyzed for percent juice content, Brix, acid, and weight. Degrees Brix and total titratable acidity were determined according to methods approved for Florida citrus quality tests (Wardowski et al., 1995).

Samples of spring growth leaves (20 leaves from five trees) and soil (two cores from each of five trees were taken from each orchard in Aug. or Sept. of each year from 1994 to 2004. Leaf samples were analyzed for N, P, K, Ca, Mg, Na, Zn, Mn, Fe, and B. Soil samples were taken at the same time to a depth of 60 cm and were analyzed for P, K, Ca, Mg, Zn, Mn, Al, Cu, Fe, Na, and Cl.

Citrus orchards in this project were irrigated with either groundwater or reclaimed water. Orchards irrigated with groundwater were managed using recommended practices receiving 30 to 40 cm of irrigation per year. However, orchards irrigated with reclaimed water had higher soil water content (Zekri & Koo, 1993), presumably because of more frequent irrigation. Orchards irrigated with reclaimed water had soil moisture content of 0.06 cm^3 cm^{-3} compared with 0.05 cm^3 cm^{-3} for orchards irrigated with ground water. Field capacity was estimated to be 0.65 cm^3 cm^{-3} for these soils, indicating that orchards irrigated with reclaimed water were near or above field capacity a higher proportion of the time compared with orchards irrigated with ground water. The quality of the reclaimed water used for irrigation was monitored monthly, and a report of average water constituent concentrations was provided to the growers (Table 1). The level of constituent concentrations in the reclaimed water are not considered to be toxic (Burton & Hook, 1979; Feigin et al., 1984). However, if soil or tissue accumulation were to occur, concentrations of heavy metals (i.e. cadmium, lead, and zinc) may approach toxic levels (Campbell et al., 1983; Feigin et al., 1984; Neilsen et al., 1991).

Prior to 1994, Zekri & Koo (1993) reported that soil to a depth of 0.5 m beneath trees irrigated with reclaimed water was usually 14.7 mm higher in water content and the trees had 6% higher canopy, leaf color, and fruit crop ratings than trees irrigated with groundwater. The higher ratings were attributed to consistently higher soil water content in the orchards irrigated with reclaimed water. For the period 1994 to 2004, mean quarterly canopy appearance, leaf color, and fruit crop, were significantly higher in orchards irrigated with reclaimed water compared with orchards irrigated with groundwater. Weed growth in orchards irrigated with reclaimed water was consistently higher, but not significantly different, than orchards irrigated with well water. The difference in mean rating for the four categories was 12.3% possibly indicating greater water use in reclaimed water blocks compared with orchards irrigated with well water.

Mean canopy, leaf color, and fruit crop ratings for trees irrigated with ground water were significantly greater than ratings from 2000 to 2004 compared with trees irrigated with the same water source from 1996 to 1999. Whereas, canopy, leaf color, and fruit crop ratings for the orchards irrigated with reclaimed water did not have a similar pattern. Reduced canopy appearance, leaf color, and fruit set in orchards irrigated with groundwater can be attributed to reduced rainfall from 1994-1999 (390 mm, 1998) compared with average rainfall from 2000 to 2004 (1191 mm). Significantly lower tree appearance in a drought year agrees with conclusions of Zekri & Koo (1993) that commercial citrus orchards irrigated with reclaimed water were commonly irrigated more frequently and/or with a greater volume than those irrigated with groundwater.

Weed growth as measured by weed cover ratings was higher in reclaimed water irrigated orchards for most years compared with those irrigated with groundwater. Higher weed growth ratings have been correlated with high irrigation rates of reclaimed water (Parsons

	Drinking water MACL	Well water typical concentrations[1]	Conserv II reclaimed water MACL	Typical Conserv II reclaimed water concentrations[1]
	-- mg L⁻¹ --------------------------------			
Arsenic	0.05	--	0.10	<0.005
Barium	2	--	1	<0.01
Beryllium	0.004	--	0.10	<0.003
Bicarbonate	--	--	200	105
Boron	--	0.02	1.0	<0.25
Cadmium	0.005	--	0.01	<0.002
Calcium	--	39	200	42
Chloride	250	15	100	75-81
Chromium	0.1	--	0.01	<0.005
Copper	1	0.03	0.20	<0.05
EC (µmhos)	781	360	1100	720
Iron	0.3	0.02	5	<0.4
Lead	0.015	--	0.1	<0.003
Magnesium	--	16	25	8.5
Manganese	0.05	0.01	0.20	<0.04
Mercury	0.002	--	0.01	<0.0002
Nickel	0.1	--	0.20	0.01
Nitrate-N	10	3	10	6.1-7.0
pH	6.5-8.5	7.8	6.5-8.4	7.1-7.2
Phosphorus	--	0.01	10	1.1
Potassium	--	6	30	11.5
Selenium	0.05	--	0.02	<0.002
Silver	0.1	--	0.05	<0.003
Sodium	160	18	70	50-70
Sulfate	250	23	100	29-55
Zinc	5	0.02	1	<0.06

[1] As reported in Parsons et al., 2001b.

Table 1. Maximum allowable contaminate limit (MACL) for Florida drinking water and Conserv II reclaimed water, and typical Water Conserv II reclaimed water concentrations. All values are in mg L⁻¹ except for pH and EC.

& Wheaton, 1992; Zekri & Koo, 1993). As with tree appearance and fruit crop, weed cover ratings only were significantly lower for orchards irrigated with groundwater in 1998 compared with other years, presumably due to lower rainfall. Growers have adjusted their herbicide practices to reduce the negative impact of increased weed growth due to higher irrigation use with reclaimed water by reducing reclaimed water use or increasing herbicide applications.

In five out of 11 years (1994, 1995, 1998, 2000, and 2001), mean fruit juice content or the percent of fruit weight in juice were significantly higher among trees in orchards irrigated with reclaimed water rather than ground water. These years with significant juice content differences among irrigation water sources lead to a significant year by water source interaction for Juice content. Juice soluble solids or Brix was not significantly different among water sources. However, Brix were significantly different among water sources in 1994, 1997 and 1998 contributing to a significant year by water source interaction. Two of these years were considered dry years with below normal rainfall. Fruit weight were significantly higher for orchards irrigated with reclaimed water compared with fruit from orchards irrigated with ground water, however, no year * water source interaction was noted. Therefore, higher fruit crop ratings, fruit weights, and similar solids per fruit (during normal rainfall years) in orchards irrigated with reclaimed water would suggest similar or greater yields in terms of soluble solids per ha compared with orchards irrigated with groundwater. The previous study by Koo & Zekri (1989) found that reduced soluble solids and acid concentration in the juice was correlated with higher soil water content in the orchards receiving reclaimed water. Likewise, significant differences in fruit Brix and acid were seen in this study from 1994 to 1998, but not after 1998. This change in fruit Brix and acid my indicated a change in irrigation practices with orchards being irrigated with similar amounts some time after 1998. This shift in irrigation practice would correspond with construction of RIBs and reduced requirement for the use of reclaimed water. Because fruit yield was greater from orchards irrigated with reclaimed water, total soluble solids produced per ha were higher in the reclaimed water orchards than the groundwater irrigated orchards.

Irrigation with reclaimed water has increased soil concentrations of P, K, Mg, B, Na, and Cl when reclaimed water was used as an irrigation water source (Burton & Hook, 1979; Campbell et al., 1983; Feigin et al., 1984; Neilson et al., 1991). Elemental concentrations in soil samples taken in Aug. or Sept. of each year from orchards irrigated with either reclaimed or ground water varied from year to year but were not significant by years. Calcium was the only element significantly different by soil sample depth with higher concentrations found near the surface. This result was expected since calcium applied as lime applied for pH adjustments in orchards irrigated with either groundwater or reclaimed water, and Ca in the reclaimed water would be incorporated into this layer with little leaching over time. With the exception of increased P, Ca and Al no elements were found to be significantly different when comparing water sources. Soil in orchards irrigated with reclaimed water was significantly higher for P, Ca and Al compared with soils in orchards irrigated with ground water. However, no elements were found to be excessive (Maurer & Davies, 1993; Tucker et al., 1995). Lower extractable soil K was found in orchards receiving higher rates of reclaimed water despite the higher K concentration of reclaimed water. These data are consistent with findings of Zekri & Koo (1993) who reported P, Ca, and Mg were significantly higher and K significantly lower in soil samples from orchards irrigated with reclaimed water compared with orchards irrigated with groundwater.

Calcium was the only element with years * water source and depth * water source interactions. Soil calcium concentrations were significantly lower (1034.7 kg ha^{-1}) in years with normal rainfall (2000-2004) compared with dryer years (1338.5, 1996-1999). Differences in soil Ca concentration among the two irrigation water sources followed the same pattern during these years with soil from orchards irrigated with reclaimed water have higher concentrations than did soil from orchards with ground water (data not shown). Likewise, soil Ca concentrations followed the same pattern with depth regardless of irrigation water source resulting in higher concentrations in soil irrigated with reclaimed water at the selected depths compared with soil from orchards irrigated with ground water.

Leaf sample elemental concentrations were generally higher from orchards irrigated with reclaimed water compared with orchards irrigated with groundwater. While higher, significantly higher P and Ca concentrations in soils irrigated with reclaimed water did not lead to significantly higher leaf concentrations. These results can be explained by dilution of leaf concentration by increased biomass production of trees irrigated with reclaimed water, reduced nutrient uptake efficiency, or a combination of the two. Unfortunately, differences in biomass accumulation were not determined in this study. However, only Mg and B were significantly higher in leaf samples from orchards irrigated with reclaimed water compared with samples from orchards irrigated with groundwater. Zekri & Koo (1993) found significantly higher Fe and B concentrations in more than half the years between 1987 and 1993. Based on this information, it is now recommended that orchards irrigated with reclaimed water not add B to micronutrients sprays. Zekri & Koo (1993) found significantly higher Na and Cl concentrations in leaf samples from orchards irrigated with reclaimed water, presumably from higher irrigation applications. However, Na and Cl were not significantly different from 1994 to 2004, further indicating a change in irrigation practice among orchards irrigated with reclaimed water.

6. Conclusion

Few detrimental effects on citrus orchards have been associated with irrigation using the reclaimed water. However, the impact of using reclaimed water on groundwater contamination have not been determined. Appearance of trees irrigated with reclaimed water was usually better, with higher canopy, leaf color, and fruit crop ratings, than orchards irrigated with groundwater. Higher weed growth in reclaimed water irrigated orchards was associated with higher soil water content. However, growers apparently have made adequate adjustments to their herbicide practices. Higher soil water content in the orchards receiving reclaimed water resulted in reduced fruit soluble solids. However, because fruit crop ratings and larger fruit size indicated greater fruit yield, total soluble solids produced per ha were similar to or higher in the reclaimed water irrigated orchards than in the groundwater irrigated orchards. Irrigation with reclaimed water generally increases soil P and Ca, and reduces soil K. Reduction of P and Ca and increases in K applied to citrus orchards irrigated with reclaimed water may be required adjustments in fertilizer applications to citrus orchards irrigated with reclaimed water. Likewise, leaf B concentration increased in most citrus trees irrigated with treated wastewater, requiring an adjustment in foliar nutrient application practices.

7. References

Al-Joloud, Hussain, A., G. Al-Saati, A.J. and Karimullah, S. 1993. Effect of wastewaters on plant growth and soil properties. Arid Soil Research Rehabilitation 7:173-179.

Allen, J.B. and McWhorter, J.C. 1970. Forage crop irrigation with oxidation pond effluent. Water Resources Research. Mississippi State.

Angelakis, A.N., Marecos De Monte, M.H.F. , Bontoux, L. and Asano, T. 1999. The status of wastewater reuse practice in the Mediterranean basin: need for guidelines. Water Resourses 33(10):2201-2217.

Arora, M. and Volutchkov, N. 1994. Water reclamation and reuse technologies commonly used in U.S.A. In Water quality International IAWQ conf. 24-29 July, 1994, pp. 343-352. Budapest, Hungary.

Asano, T., and Levine, A.D. 1996. Wastewater reclamation, recycling and reuse: pas, present and future. Water Sci. Technol. 33(10):1-6.

Banuls, J., Legaz, F. and Primo-Millo, E. 1990. Effect of salinity on uptake and distribution of chloride and sodium in some citrus scion-rootstock combinations. J. Hort. Sci. 65:722-724.

Basiouny, F.M. 1982. Wastewater irrigation of fruit tree. Biocycle 23(2): 51-53.

Basiouny, F.M. 1984. The use of municipal treated effluent for peach tree irrigation. Proc. Fl. State Hort. Soc. 97:345-347.

Berry, W.L., Wallace, A. and Lunt, O.R. 1980. Utilization of municipal wastewater for culture of horticultural crops. HortScience 15:169-171.

Bielorai, H., Vaisman, I. and Feigini, A. 1984. Drip irrigation of cotton with treated municipal effluents: I. Yield response. J. Envir. Qual. 13:231-234.

Bole, J.B., and Bell, R.G. 1978. Land application of municipal wastewater: Yield and chemical composition of forage crops. J. Envir. Qual. 7:222-226.

Burton, T.M. and Hook, J.E. 1979. A mass balance study of application of municipal wastewater to forest in Michigan. J. Environ. Qual. 8:589-596.

Campbell, W.F., Miller, R.W. Reynolds, J.H. and Schreeg, T.M. 1983. Alfalfa, sweet corn, and wheat responses to long-term application of municipal wastewater to cropland. J. Environ. Qual. 12:243-249.

Cramer, G.R., Lynch, J. Lauchli, A. and Epstein, E. 1987. Influx of Na, K, and Ca into roots of salt-stressed cotton seedlings: effects of supplemental Ca. Plant Physiol 83:510-516.

Crites, R. 1975. Wastewater irrigation. Waterwaste Eng. 12:49-50.

Day, A., Take, F. and Katterman, F. 1975. Effects of treated municipal wastewater on growth, fiber, acid soluble nucleosides, protein and amino acid content in wheat grain. J. Environ. Qual. 4:167-169.

Day, A. and Tucker, C.T. 1977. Effects of treated municipal wastewater on growth, fiber, protein, and amino acid content of sorghum grain. J. Environ. Qual. 6:325-327.

Day E.O. 1958. Crop irrigation with sewage effluent. Sewage and Industrial Wastes 30:825-828.

Embleton, T.W., Jones, W.W. Labanauskas, K. and Teuther, W. 1973. Leaf analysis as a diagnostic tool and guide to fertilization. In: Reuther W. (ed) The citrus industry. University of California. Berkeley p 183.

Epstein, E. 1961. The essential role of calcium in selective cation transport by plant cells. Plant Physiol. 36:437-444.

Esteller, M.V., Duran, A., Morell, I., Garcia-Agustin, P. and Lapena, L. 1994. Experimental citrus irrigation with reclaimed wastewater on a Spanish coastal aquifer. In: Reeve, C and W. Watts (eds) Groundwater, drought pollution and management. Rotteram. P 55.

Feigin, A., Vaisman, I. and Bielorai, H. 1984. Drip irrigation of cotton with treated municipal effluents. II. Nutrient availability in soil. J. Environ. Qual. 13:234-238.

Feigin, A., Ravina, I. and Shalhevet, J. 1990. Irrigation with treated sewage effluent. Ecological Series, Springer Verlag, New York.

Florida Department of Environmental Protection. 2005. Reuse inventory report. Available at www.dep.state.fl.us/water/reuse/inventory.htm (verified 09/28/2006)

Hanlon, E.A. andDeVore, J.M. 1989. IFAS extension soil testing laboratory chemical procedures and training manual. Inst. Food Agric. Sci., Univ. of Fla., Gainesville, Circ. 812.

Haruvy, N., and Sadan, E. 1994. Cost benefit analysis of wastewater treatment in the water scarce economy of Israel: a case study. J. Financial Management and Analysis 7(1):44-51.

Haruvy, N. 1994. Recycled water utilization in Isreal: focus on wastewater pricing versus national and farmer objectives. J. Financial Management and Analysis 7(2):39-49.

Haruvy, N. 1997. Agricultural reuse of wastewater: nation-wide cost-benefit analysis. Agriculture, Ecosystems and Environment 66:113-119.

Hass, C.N., Rose, J.B., Gerba, C. and Regli, S. 1993. Risk Assessment of virus in drinking water. Rick Analysis, 13:545-552.

Henry, C., Maldenhauer, R., Engbert, L. and Troug, E. 1954. Sewage effluent disposal through crop irrigation. Sewater and Industrial Wastes 26:123-133.

Kale, C.K. and Bal, A.S. 1987. Reuse of stabilization pond effluent for citrus reticulate (orange), forest and road verge plants. Wat. Sci. Tech. 19(12):307-315.

Karlen, D.L., Vitosh, M.L. and Kunze, R. 1976. Irrigation of corn simulated municipal sewage effluent. J. Environ. Qual. 5:269-273.

Kirkham, M.B. 1986. Problems of using wastewater on vegetable crops. HortScience 21:24-27.

Koo, R.C.J. 1963. Effects of frequency of irrigation on yield of orange and grapefruit. Proc. Fla. State Hortic. Soc. 74:1-5.

Koo, R.C.J. and McCornack, A.A. 1965. Effects of irrigation and fertilization on production and quality of 'Dancy' tangerine. Proc. Fla. State Hort. Cos. 78:10-15.

Koo, R.C.J. and Smajstrla, A.G. 1984. Effects of trickle irrigation and fertigation on fruit production and juice quality of 'Valencia' orange. Proc. Fla. State Hort. Soc. 97:8-10.

Koo, R.C.J. and Zekri, M. 1989. Citrus irrigation with reclaimed municipal wastewater. Proc. Fla. State Hortic. Soc. 102:51-56.

LaHaye, P.A., andEpstein, E. 1969. Salt toleration by plants: enhancement with calcium. Science 166:395-396.

Lapena, L., Cerezo, M.and Garcia-Augustin, P. 1995. Possible reuse of treated municipal wastewater for citrus spp. plant irrigation. Toxicol. 55:697-703.

Maas, E.V. and Hoffman, G.J. 1977. Crop salt tolerance current assessment. ASCE J Irri. Drainage Division 103:115-134.

Marella, R.L. and Berndt, M.P. 2005. Water withdrawal and trends from the Floridan aquifer system in the southeastern United States, 1950-2000. USGS Circ. 1278. Washington, DC.

Maurer, M.A. and Davies, F.S. 1993. Microsprinkler irrigation of young 'Redblush' grapefruit trees using reclaimed water. HortScience 28(12):1157-1161.

McMahon, B.R.R., Koo, R.C.J. and Persons, H.W. 1989. Citrus irrigation with reclaimed wastewater. Trans. Citrus Engr. Conf. 35:1-17.

Morgan, K.T., Wheaton, T.A. Parsons, L.R. and Castle, W.S. 2008. Effects of reclaimed municipal waste water on horticultural ratings, fruit quality, and soil and leaf mineral content of citrus grown in central Florida. HortScience 43(2):459-464.

Morvedt. J. and Giovdane, P. 1975. Response of corn to zinc and chromium in municipal waste applied to soil. J. Environ. Qual. 4:361-366.

Neilsen, G.H., Stevenson, D.S. and Fitzpatrick, J.J. 1989a. The effect of municipal wastewater irrigation and rate of N fertilization on petiole composition, yield, and quality of Okanalgran Riesling grapes. Can. J. Plant Sci. 69:1285-1294.

Neilsen, G.H., Stevenson, D.S., Fitzpatrick, J.J. and Brownlee, C.H. 1989b. Nutrition and yield of young apple trees irrigated with municipal waste water. J. Amer. Soc. Hort. Sci. 114(3):377-383.

Neilsen, G.H., Stevenson, D.S., Fitzpatrick, J.J. and Brownlee, C.H. 1989c. Yield and plant nutrient content of vegetables trickle-irrigated with municipal wastewater. HortScience 24:249-252.

Neilsen, G.H., Stevenson, D.S., Fitzpatrick, J.J. and Brownlee, C.H. 1991. Soil and sweet cherry responses to irrigation with wastewater. Can. J. Soil Sci. Soc. 71:31-41.

Nguy, O.C., 1974. Yield response and nitrogen uptake by forage crops under sprinkler irrigation with treated municipal wastewater. M.X. Thesis, Unvi. Florida, Gainesville.

Obreza, T.A. and Collins, M.E. 2003. Common soils used for citrus production in Florida, Univ. of Fla., Gainesville, Bull. SS403.

Obreza, T.A. and Morgan, K.T. 2008. Nutrition of Florida citrus trees. University of Florida, Gainesville, Bull. SL253.

Omran, M.S., Waly, T.M., Abd Elnaim, E.M. and El Nashar, B.M.B.. 1988. Effect of sewage irrigation on yield, tree components and heavy metals accumulation in navel orange trees. Biological Wastes 23:17-24.

Parsons, L.R. and Wheaton, T.A.. 1992. Reclaimed water — a viable source of irrigation water for citrus. Proc. Plant Stress Tropical Environ. 25-26.

Parsons, L.R., Wheaton, T.A. and Cross, P. 1995. Reclaimed municipal water for citrus irrigation in Florida. P. 262-268. In: F.R. Lamm (ed) Microirrigation for a changing world. Proc. Fifth Intl. Microirr. Congr. Amer. Soc. Agr. Eng.

Parsons, L.R. and Wheaton, T.A.. 1996. Florida citrus irrigation with municipal reclaimed water. Proc. Intl. Soc. Citricult. 692-695.

Parsons, L.R., Morgan, K.T. and Wheaton, T.A. 2001a. Wastewater and reclaimed water — disposal problem or potential resource? Proc. Fla. State Hortic. Soc. 114:97-100.

Parsons, L.R., Wheaton, T.A. and Castle, W.S. 2001b. High application rates of reclaimed water benefit citrus tree growth and fruit production. HortScience 36:1273-1277.

Perry, M.J. and Mackum, P.J. 2001. U.S. population changes and distributions 1990 to 2000. U.S. Census Bureau, U.S. Dept. of Commerce, Washington, DC. Available at http://www.census.gov/prod/2001pubs/c2kbr01-2.pdf#search=%22Florida%20population%202000%20growth%22 (verified 09/28/2006)

Pettygrove, G.S. and Asano, T. 1985. Irrigation with reclaimed municipal wastewater - a guidance manual. Lewis Publishers Inc. Chelsea, Michigan.

Ramos,C., Gomez de Barred, D. Oliver, J. Lorenzo, E.and Castell, J.R. 1989. Aguas residuals para riego: un Ejemplo de aplicacion en uva de mesa. In: Cabrera, E. and A. Sahuquillo. (eds) El agua en la Comunidad Valenciana. Generalitat Valenciana, Spain p 167.

Regli, S., Rose, J.B., Hass, C.W. and Gerba, C.P.1991. Modeling risk for pathogens in drinking water. J. Am. Water Works Association 83:76-84.

Sanderson, K.C. 1986. Introduction to the workshop on wastewater utilization in horticulture. Hortscience 21(1):23-24.

Shuval, H., Lampert, Y. and Fattal, B.. 1997. Development of a risk assessment approach for evaluating wastewater reuse standards for agriculture. Wat. Sci. Tech. 35(11):15-20.

Singh, K.P., Mohan, D., Sinha, S. and Dalwani, R. 2004. Impact assessment of treated/untreated wastewater toxicants discharged by sewage treatment plants on health agricultural and environmental quality in the wastewater disposal area. Chemosphere 55:227-255.

Smith, S.K. 2005. Florida population growth: past, present, and future. University of Florida, Bureau of Economic and Business Research Report. Available at http://www.bebr.ufl.edu/Articles/FloridaPop2005.pdf#search=%22Florida%20population%201950%22 (verified 09/28/2006)

State of California. 1978. Wastewater reclamation criteria, an excerpt from the California code of regulations. Title 22, Division 4 Environmental Health, Department of Health Services, Sacramento, California.

Stokes, W.E., Leukel, W. and Bannette, R. 1930. Effect of irrigation with sewage effluent on the yield and establishment of Napier grass and Japanese cane. J. Amer. Soc. Hort. Sci. 61:38-48.

Tucker, D.P.H., Alva, A.K., Jackson, L.K. and Wheaton, T.A. 1995. Nutrition of Florida citrus trees. Univ. of Florida. Coop. Ext. Serv. SP169.

U.S. Census Bureau. 1997. Florida population by counties. Dept. of Commerce, Washington, D.C. Available at http://www.census.gov/population/cencounts/fl190090.txt (verified 09/28/2006)

USEPA. 1992. Guidelines for water reuse. US Environmental Protection Agency (Technical Report 81), Washington D.C. 225 pp.

Walker, R.R., Torokfalvy, E. and Downton, W.J.S. 1982. Photosynthetic responses of the Citrus varieties Rangpur lime and Etog citron to salt treatment. Aus J Plant Physiol. 9:783-790.

Wardowski, W., Whigham, J., Grierson, W.andSoule, J. 1995. Quality tests for Florida citrus. Univ. of Fla., Gainesville, Bull. SP 99.

Wheaton, T.A. and Parsons, L.R. 1993. Reclaimed water for citrus: Effects of irrigation rate on tree growth and yield. HortScience 28(5):497.

World Health Oranization. 1989. Health Guidelines for the use of wastewater in Agriculture and Aquaculture. Peprot of the World Health Organization Scientific Group (Technical Report Series 778, Geneva, 74 pp.

Zekri, M. and Koo, R.C.J. 1990. Effect of reclaimed wastewater on leaf and soil mineral composition and fruit quality of citrus. Proc. Fla. State Hort. Soc. 103:38-41.

Zekri, M. and Koo, R.C.J. 1993. A reclaimed water citrus irrigation project. Proc. Fla. State Hortic. Soc. 106:30-35.

Zekri, M. and Koo, R.C.J. 1994. Treated municipal wastewater for citrus irrigation. J Plant Nutrition 17:693-708

5

Waste Water: Treatment Options and its Associated Benefits

Akuzuo Ofoefule, Eunice Uzodinma and Cynthia Ibeto
Biomass Unit, National Center for Energy Research & Development,
University of Nigeria, Nsukka. Enugu state
Nigeria

1. Introduction

Wastewater, a semi-liquid waste that is discharged from residential homes, industries, agricultural and commercial properties potentially release significant amounts of toxic and pathogenic contaminants into local treatment plants for processing. These contaminants include not only soaps, shampoos and conditioners used in showers, food scraps and oils from kitchen sinks and garbage disposals, human waste from toilets, detergents and fabric softeners from washing machines and dishwashers as well as all the harsh detergents that clean the house, but also heavy metals, pharmaceuticals, volatile organic compounds (VOCs), volatile toxic organic compounds (VTOCs), pathogenic microorganisms, phosphorus, nitrogen, substances that are carcinogenic, tetragenic and mutagenic that are resistant to typical wastewater treatment processes that come from industries.

Potable water becomes wastewater after it gets contaminated with natural or synthetic microbiological compounds that arise out of human activities, commercial and industrial sources. They may be accompanied with surface water, ground water and storm water. Wastewater is also sewage, storm-water and water that have been used for various purposes around the community. Unless properly treated, wastewater can be harmful to public health and the environment.

2. Sources of wastes

Most communities generate wastewater from both residential and non-residential sources.

Residential Wastewater or Household Wastewater

Residential wastewater is a combination of excreta, flush water and all types of wastewater generated from every room in a house. It is more commonly known as sewage and is much diluted. There are two types of domestic sewage: black-water or wastewater from toilets, and gray water, which is wastewater from all sources except toilets. Black-water and gray-water have different characteristics, but both contain pollutants and disease-causing agents. In the U.S, sewage varies regionally and from home to home. These are based on factors such as the number and type of water-using fixtures and appliances used at homes and even their habits, such as the types of food that are eaten.

Non-Residential Wastewater or Industrial Wastewater

This is mainly made up of wastes coming from commercial activities (e.g., shops, restaurants, hospitals etc.), Industry (e.g., Chemical Industries, Pharmaceutical companies, Textile manufacturing companies etc.), Agriculture (e.g., slurry), construction and demolition projects, mining and quarrying activities and from the generation of energy. These could be places such as industrial complexes, factories, offices, restaurants, farms and hospitals. Because of the different non-residential wastewater characteristics, communities need to assess each source individually to ensure that adequate treatment is provided. For example, laundries differ from many other industrial sources because they produce high volumes of wastewater containing lint fibers. Restaurants typically generate a lot of oil and grease. In addition, many industries produce wastewater high in chemical and biological pollutants that, can overburden onsite and community wastewater treatment systems.

Storm-water is a nonresidential source and carries trash and other pollutants from streets, as well as pesticides and fertilizers from yards and fields. Communities may require these types of nonresidential sources to provide preliminary treatment to protect community systems and public health (Runion, 2010).

Environmental hazards of waste water

Wastewater can attract rodents and insects which cause gastrointestinal parasites, yellow fever, worms, the plague and other unhealthy conditions for humans. Exposure to hazardous wastes, particularly when they are burned, can cause various other diseases including cancers. Wastes can contaminate surface water, groundwater, soil, and air which causes more problems for humans, other species, and ecosystems. Waste treatment and disposal produces significant green house gas (GHG) emissions, notably methane, which are contributing significantly to global climate change.

Disease- causing pathogens are constantly being released into waterways from waste water. However, these chemical substances are only the tip of the iceberg. The pathogens from diseases such as AIDS, cholera, plague, hepatitis, typhoid, polio and so on are also released from homes, medical clinics, laboratories and hospitals. Studies have shown that every gram of fecal material from an infectious hepatitis patient can contain up to 100,000 infectious doses. Other pathogens include *cryptosporidium, giardia, neospora, e.coli, stretococcus, legionella, salmonella, shigella, vibrio,* adenoviruses, Norwalk, rotavirus, amoeba, whipworm, tapeworms, flukes, pinworms, roundworms, *klebsiella, clostridium, pseudomonas* and *mycobacterium tuberculosis.* These microbes are not looked for nor tested in a routine analysis of treated wastewater before their release into the environment.

Many viruses can survive in wastewater up to 41 days at 20°C. Once released into the environment, they can survive up to six or more days in a river and up to 100 days in soil. The protozoa parasite can survive up to 20 days in soil while bacteria can survive up to 120 days. Most worms like the ascaris, tapeworms and trichuris can survive up to 12 months in soil. Their survival in soil depends on moisture, pH, temperature, type of soil and the presence of organic matter (Anon, 1980). It is estimated that every year 1.8 million people die worldwide due to suffering from waterborne diseases. A large part of these deaths can be indirectly attributed to improper sanitation. Wastewater treatment is an important initiative which has to be taken more seriously for the preservation of society both at present and in the future.

Also, Mills discharge millions of gallons of effluent each year, full of chemicals such as formaldehyde (HCHO), chlorine, heavy metals (such as lead and mercury) and others, which are significant causes of environmental degradation and human illnesses. The mill effluent is also often of a high temperature and pH, both of which are extremely damaging. All of the mills O Ecotextiles (A producer of high quality organic fabrics in Seattle, Washington) uses, have wastewater treatment in place. Every 25 meters of an O Ecotextiles sofa fabric prevents 2,300 liters of chemically infused effluent(about the size of a California hot tub and containing from 1 to 10 kg of toxic chemicals), from entering the environment (Based on VPI study for Dept. of Environmental Quality for the state of Virginia.) (Anon, 2005).

Some advantages of waste water and its treatment

Oboh (2005) studied the utilization of fermented waste water from cassava mash as source of industrial amylase and reported that the amylases from fermented cassava waste water are active at wide temperature and pH ranges. This quality could be explored in the industrial sector (most especially food industry) as a source of industrial amylase.

Wastewater treatment is a process whereby the contaminants are removed from wastewater as well as household sewage, to produce waste stream or solid waste suitable for discharge or re-use (Naik, 2010). Treated wastewater is now being considered as a new source of water that can be used for different purposes such as agricultural (70% of Israel's irrigated agriculture is based on highly purified wastewater) and aquaculture production, industrial uses (cooling towers), recreational purposes and artificial recharge. Using wastewater for agricultural production will help in alleviating food shortages and reduce the gap between supply and demand. Treated wastewater can be re-used as drinking water, in artificial recharge of aquifers, in agriculture and in the rehabilitation of natural ecosystems (Florida's Everglades).

Re-use of wastewater, in concert with other water conservation strategies, can help lessen anthropogenic stresses arising from over-extraction and pollution of receiving waters. However, there are concomitant environmental risks with wastewater re-use, such as transport of harmful contaminants in soils, pollution of groundwater and surface water, degradation of soil quality e.g. salinization, impacts on plant growth, the transmission of disease via the consumption of wastewater-irrigated vegetables, and even increased greenhouse gas emissions. The challenge facing wastewater re-use is to minimize such risks so as to maximize the net environmental gain.

There are more than 150 known pathogens detected in untreated wastewater. Every year new ones are being discovered. Of the 72 enteroviruses, many will trigger illnesses that are not gastrointestinal, such as, polio, meningitis, diabetes, muscle diseases and endocarditis (inflammation of the heart muscle that can lead to heart attacks). They can and do produce infectious illnesses in humans that multiply and are re-excreted through fecal material (Mara and Horan, 2003), hence, the need for waste water treatment in order to avoid the occurrence of such pathogens in the environment.

Benefits of treatment on man and the environment

Endocrine disruptors, also known as xenoestrogens, are chemical compounds and by-products used in the plastic, pesticide and chemical industries and found in their waste water that have hormonal effects on the body. There are more than 100,000 registered endocrine disruptors. They are far more potent in mimicking estrogen activity than the body's natural hormones and far more toxic. The synergistic effects of these chemicals in the body may be up to 1000 times greater. Endocrine disruptors create a large range of reproductive problems. They include infertility, menstrual problems, difficulty holding a pregnancy to term and early puberty. Other health issues include impaired immune function, behavioral problems, brain malfunctions and cancer (Anon, 1995).

Waste water quality deteriorates rapidly and causes microorganisms, protozoa, fecal *coliform*, and fecal *streptococci* in the surrounding area. It is affected directly by the quality of groundwater in the area mainly with surfactants such as Anionic Detergents (linear alkyl sulphonate, LAS) and nitrite in the surrounding area around the infiltration basins with cycle diameter of 800 m. This is very harmful to the human body and cause gastroenteritis troubles.

Wastewater Treatment

Wastewater treatment is the removal of solids, bacteria, algae, plants, inorganic and organic compounds from used water with subsequent conversion into environmentally acceptable water or even drinking water. This treatment usually requires science, engineering, business and art (Anon., 2010) and includes mechanical, physical, chemical and biological methods. Most often, huge amount of waste water is generated from various sources; domestic, municipal and industrial activities. The waste water may be characterized with pH, suspended solids (SS) dissolved solid (DS), turbidity, colour, biological oxygen demand (BOD) and chemical oxygen demand (COD) among others. Each source of the water has its own pollution problem and must be treated to return water to its natural state or re-used in various activities. Waste water treatment can be grouped in three major ways: physical, chemical and biological treatment. Physical treatment involves processes such as sedimentation, aeration, filtration and floatation while chemical treatment uses oxidizing agents such as chlorine, ozone, including neutralization method. Sometimes coagulants like alum, iron (III) sulphate, among others could be utilized. Carbon could be used both in chemical and physical methods as adsorbent. In the biological water treatment, microorganisms such as bacteria are biochemically employed to decompose wastewater to produce water stable product. There are aerobic and anaerobic decomposition. Under aerobic decomposition, contaminants in water are converted to carbon dioxide while anaerobic decomposition produces methane as a fuel in biogas and carbon dioxide as major products. The effluent from anaerobic process can be used as manure in agricultural production. (Anon, 2010).

The Importance/need for Waste water Treatment

Population explosion, disorderly rapid movement to the urban areas, technological and industrial expansion, energy utilization and waste generation from domestic, municipal and industrial sources have rendered many waters unwholesome and hazardous to man and other living things. There are no stringent laws guiding environmental pollution in most developing countries in the world. Consequently, industries discharge untreated or inadequately treated water into water ways (Amuda and Ibrahim, 2006). Most of these waters pose great dangers to aquatic life and must be cleaned or treated.

Reported Researches on Wastewater Treatment Methods

Treatment processes for heavy metals removal from wastewater include precipitation, membrane filtration, ion exchange, adsorption and co- precipitation/adsorption. Studies have shown that adsorption method is highly effective and activated carbon has been in use but very expensive (Nasim *et al.*, 2004). Amuda and Ibrahim (2006), reported on the comparison of adsorption efficiency of coconut shell- based activated carbon and commercial activated carbon calgon, carbon (F-300) in the treatment of industrial wastewater from a beverage industry for the removal of organic matter (COD). The results of their findings indicated that coconut shell- based granular activated carbon was found to effectively adsorb the organic than the commercial activated carbon. Besides, previous researches on the removal of COD from industrial wastewater using coagulation/flocculation, membrane filtration and oxidation processes etc., reveal that the technology of these methods are generally expensive, complicated, time consuming and required skilled personnel (Galambos et al., 2004; Peres et al., 2004). However, low cost and non-conventional adsorbents such as nut shells, wood, bone, peat, processed into activated carbons and biomass such as Aspergillum tereus, Peusodomonas sp., Rhizopus arrhizus are better (Okieimen et al., 1985; Tam and Antal, 1999; Bansode et al.,2004; Nomanbhay and Palanisamy, 2005; Preetha and Viruthagiri, 2005). Again, ultraviolet (UV) lamp has also been used to treat textile wastewater. Stanislaw and Monica (1999), reported the application two different UV radiations; 150W, λ=254-578nm and 15W, λ=254nm to the synthetic textile wastewater for 1-3h. There was significant inhibitory action on the microbes (47 to 30% reductions) for the optimum radiation time. Moreover, ozone (O_3) is a powerful oxidant for both water and wastewater. The conventional fine bubble contactor is the most widely used Ozone generator because of high Ozone transfer efficiency (90%). Result of findings from Mehmet and Hassan (2002), showed that ozonation using $300mg/dm^3$ gave rise to biodegradability index of textile wastewater by 1.6times. Few researchers namely; Namboodri et al., (1994); Rajeswari (2000) among others revealed that ozone decolourizes all dyes except non-soluble dispersed and vat dyes that react slowly and take longer time. Hence, ozone was combined with other agents for its complete oxidation of organic compounds in the wastewater to carbon dioxide and water. This combination has led to advanced oxidation processes using ozone and others, e.g.---O_3/UV, O_3/H_2O_2, O_3/TiO_2, etc. Xuejun et al., (2005), used a combination of electrochemical oxidation combined with membrane filtration to treat wastewater from a textile industry. Results obtained showed that electrochemical oxidation has a high removal of 89.9% efficiency of COD of wastewater whereas the membrane filter can almost completely remove total suspended solids (about 100%) and turbidity (98.3%). After electrochemical and membrane filtration steps were employed, COD levels were reduced to 18.2mg/L from 178.5mg/L. Turbidity was reduced from 18.5 NTU to 0.32 NTU. Oparaku et al., (2011) reported on disinfection of wastewater from fish pond for re-utilization using ultraviolet radiations. The UV treated wastewater had lowest coliform count of $1.8X10^3$ Cfu/ml in comparison with solar and electric powered pump water counts obtained as $2.2X10^3$ and $6.8X10^3$ Cfu/ml, respectively. This treated water also had dissolved oxygen that amounted to 7.2mg/L, averagely favourable for aquatic life. Moreover, anaerobic digestion has been utilized to convert municipal wastewater (MSW) into methane and carbon dioxide with the effluent used as biofertilizer. Chynoweth et al., (1991), presented the results of biochemical potential of several fractions of MSW in order to compare the extent and rates of their conversions to methane. It was discovered that the methane yield was as high as $0.20m^3/kg$ of volatile solids added. A report from Okafor (1998) showed that cassava wastewater generated from "garri" production (a staple food consumed in the Eastern part of Nigeria), was inoculated with microorganisms to produce microbial biomass. This biomass was then mixed with ground cassava peels to formulate feed for pigs. Droppings from the pigs were later converted to biogas through anaerobic digestion. Again, implementation of anaerobic process on wastewater from tapioca starch industries has been reported. The research results showed that the value of organic substance in the influent were in the average of 10, 062 and 5,649 ppm in terms of COD and BOD, respectively. Maximum organic loading rate applied was 7.8kg COD/m^3 day. The efficiency of degradation reached an average of 76% and 95.8% for COD and BOD, respectively. Methane content in the biogas was in the range of 53.5 to 71% while average biogas production was $1.2m^3/m^3$ of wastewater.

3. Anaerobic digestion as a waste water treatment option

Anaerobic digestion (AD) is biological process similar in many ways to composting. It is a natural treatment process and as in composting bacteria in the absence of air, breakdown organic matter and reduce its bulk or mass (polymers) into simpler compounds mainly methane (CH_4) and carbon IV oxide (CO_2) and traces of other gases like O_2, H_2S, NH_3, N_2, CO and water vapour etc (Wolfgang and Axel, 2005). The effluent of this process is a residue rich in essential inorganic elements like nitrogen and phosphorus needed for healthy plant growth known as biofertilizer which when applied to the soil enriches it with no detrimental effects on the environment (Bhat et al., 2001).

Anaerobic waste digestion Takes place in a closed reactor. Bacteria act upon the organic waste and release a lot of methane and carbon dioxide. The microbial community has only obligate anaerobic and facultative bacteria. As in aerobic chemohetrotrophic metabolism, initially the macromolecules are hydrolyzed. These products are then converted to volatile fatty acids (mainly acetic acid), and alcohols. The organisms responsible for these anaerobic waste reactions are popularly called acid formers or acidogens. They obtain energy through oxidation of organic compounds, but do not use oxygen as electron acceptor. Instead, another fragment of the substrate is reduced to anaerobic acids and alcohols through anaerobic processes. These are then metabolized by a second group of obligate anaerobic biomass (the methane formers or methanogens), and are converted to methane gas. It is estimated that 60 to 70% of methane production in an anaerobic waste reaction is through conversion of acetic acid and the rest through carbon dioxide reduction by hydrogen. The activities of the methane and acid- producing groups of bacteria must be balanced as the former is sensitive to pH changes and works best in pH range 6.8 to 7.5. (Runion, 2010).

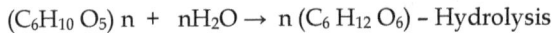

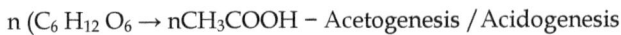

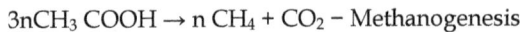

$$(C_6H_{10}O_5)\,n \;+\; nH_2O \rightarrow n\,(C_6\,H_{12}\,O_6) - \text{Hydrolysis}$$

$$n\,(C_6\,H_{12}\,O_6 \rightarrow nCH_3COOH - \text{Acetogenesis / Acidogenesis}$$

$$3nCH_3\,COOH \rightarrow n\,CH_4 + CO_2 - \text{Methanogenesis}$$

The objectives for planning an anaerobic digestion process include

1. To provide waste regulations compliant, sustainable and cost effective method of disposal of organic wastes
2. To treat, clean source separated organic wastes from households, restaurants, industries and other enterprises in an environmentally friendly manner.
3. To provide a sustainable and cost effective method of disposal of ABPR waste materials
4. To reduce carbon emissions and carbon levies payable associated with a business.
5. To establish a sustainable circulation of plant nutrients and organic material between the community and agricultural sector in such a way that the use of the residual is optimized.
6. To provide opportunities for the use of artificial fertilizer (i.e. promote organic farming).
7. To extract and use high grade bio-energy from waste and normal farm crops, with no net contribution to the atmosphere.
8. To promote and develop high efficiency energy processes and remove odours generated from present systems.
9. To reduce risk of water pollution from current practices and generally reduce emissions of enteric organisms and water courses.

Environmental technologists have in the last decade been concerned with straight forward technological, economic challenges such as drinking water production, waste water treatment, refuse handling and treatment , soil and sediment clean-up and waste gas purification. Only recently, they have started to look at their activities from the point of view of sustainability and they have had to admit that in many cases, they were far from holistic (Verstraete and Top, 1992).

Typical examples of non-sustainable approaches are current practices in anaerobic waste water treatment and refuse land filling (Allison-Onyechere et al., 2007). Lettinga et al., 1980 and Verstraete et al., 1996 reported the development of anaerobic sludge blanket or UASB reactor for the treatment of waste water. In this process, the waste water is pumped upwards through a reactor under strictly anaerobic conditions at a rate between 0.5 and 1.5m/hr. Inside the reactor, a selection process occurs which can result in the growth of anaerobic micro-organisms in kind of conglomerate (granules) varying between 0.5-5mg in diameter. These granules are powerful biocatalysts and convert the biodegradable organic matter in the influent in a rapid (space loadings varying from 10-20kg chemical oxygen demand (COD)/m^3/reactor/day) and complete way to biogas. Actually, the granular biomass is such a valuable biocatalyst that it is the only mixed culture which is commercially handled world wide at a respectful price of the order of about 1-2USD per Key dry weight. The sludge is separated from the water and the gas phase by means of an internal settler. Generally effluents approaching discharge standards are thus obtained from waste waters from breweries and soft drink plants, from potato processing plants and from certain paper recycling plants.

For concentrated waste waters, an aerobic treatment has to succeed the anaerobic treatment, yet in a similar and less energy consuming manner. At present, several hundreds of UASB reactors have been installed worldwide, particularly to treat industrial waste water with a COD exceeding 2.0 grams per litre (g/l). They are generally implemented when the waste water is rich in carbohydrates and relatively poor in other contaminants (Allison-Onyechere et al., 2007). For tropical countries, direct anaerobic treatment of sewage has become a reality and several full scale systems are currently operated in Columbia, Brazil and India (Van Haandel and Lettinga, 1994). This is a channel that must be explored further in terms of additional sun-heat input (e.g by means of solar heating systems) and nutrient removal by means of nutrient immobilizing straw biofilter as reported by Avnimelech et al., 1993.

4. Anaerobic digestion of water slurries

The authors have carried out anaerobic digestion of different waste waters / slurries and they are highlighted below;

5. Methodology

Wastes procurement

The wastes used for the anaerobic digestion studies discussed below, which invariably were the wastes to be treated were generally procured or collected from the locality. For instance, the cassava waste waters were collected from local processors of "garri" (a staple food consumed in the eastern part of Nigeria). The palm oil sludge was collected from local processors of palm oil in the community. The cow liquor waste was collected from an abattoir in the locality while the Brewery spent grain and carbonated soft drink sludge were obtained from Nigerian Breweries Ltd and 7UP bottling company (a soft drink manufacturing company) all around the locality. The swine dung were collected from the Veterinary farm in the University of Nigeria, Nsukka.

Waste preparation

Since the primary focus of the different studies were on cheaper sources of energy generation while secondarily obtaining waste effluents that are not only harmless to the environment but expected to boost organic farming/Agriculture, some of the waste waters were pretreated with other solid organic wastes to improve on their biogas production potentials. For instance,

1. In one study, cassava waste water (CW) was co- digested with swine dung (SD) in the ratio of 2:1 while the CW alone and SD alone served as control (Ofoefule et al., 2010).
2. In another study, Abattoir cow liquor (CLW) was combined with brewery spent grain (BS) in the ratio 1:1, cassava waste water (CW) in the ratio 1:3 and carbonated soft drink sludge(CS) in the ratio 3:1 (Uzodinma and Ofoefule, 2008).
3. In yet another study, palm oil sludge (POS) was blended with Agro-industrial wastes like cassava waste water (CW), Brewery spent grain (SG) and carbonated soft drink sludge (SL) in ratios of 1:1, 1:1, and 1:1.2 respectively (Uzodinma et al., 2007a).
4. In yet another study, cassava waste water (CW) was combined with palm oil sludge (POS) in the ratio 2:1, powdered rice husk (RH) 1:2.3 and pig dung (PD) 1:1.5. The nature of the wastes determined their combination ratios (Uzodinma et al., 2007b).

Non- waste materials

Other materials used in the studies generally included; metallic prototype biodigesters/fermenters ranging from 50L capacity (Fig. 1) to 136L capacity (Fig. 2) fabricated locally at the National Centre for Energy Research and development, University of Nigeria, Nsukka. Other materials also used were; Top loading balance (50kg capacity "Five goats" model no Z051599), plastic water troughs, graduated transparent plastic buckets for measuring daily gas production, digital pH meter (Jenway 3510), thermometer (-10-110°C), hosepipes, biogas burner fabricated locally for checking gas flammability.

Fig. 1. 50L Capacity Metallic Prototype Biodigester

Fig. 2. 136L Capacity Metallic Prototype Biodigester

Experimental studies

The wastes were generally mixed with water in the ratio of 2:1 except in the cases where the wastewaters were used alone as control. In such instances, the waste waters were used as they were without further dilution since the constituents were mainly water (93-95%). The digesters were charged up to ¾ level leaving ¼ head space for gas collection. They were stirred thoroughly and on a daily basis throughout the retention period to ensure homogenous blend of the wastes and dispersion of microbes in the entire mixture. Gas production measured as dm^3/kg slurry or L/Total mass of slurry were obtained by downward displacement of water by the gas.

Analyses of wastes

Physicochemical properties of the wastes such as ash, moisture, crude fibre contents, crude fat, crude nitrogen and protein contents, carbon, energy, total and volatile solids were generally determined for all the wastes using recognized laboratory procedures. These properties inherent in the wastes determine and explain the behavior of the wastes during anaerobic digestion. Biochemical analyses such as pH, ambient and influent temperatures were also monitored on the waste slurries as the digestion of the waste progressed. Microbial analysis was also carried out to determine the microbial total viable counts (TVC) for the waste slurries at different periods during the digestion; at the point of charging the digester, at the point of flammability, at the peak of gas production and at the end of the retention period. In some cases flammable gas composition from the different wastes were also analyzed.

6. Results and discussion

The various results obtained during each of the studies are as itemized below:

1. **Anaerobic batch co-digestion of cassava waste water and Swine dung**

The cassava waste water alone had the highest yield of biogas production (130 dm^3/Total mass of slurry) even though the gas produced was not flammable throughout the retention period and therefore does not meet the desired need for cooking and lighting but would however be okay for the purposes of ordinary treatment of the waste water. The non flammability of the gas produced was attributed to the acidic nature of the waste. The microbes that convert wastes to biogas are pH sensitive and survive optimally within the pH range of 6.5-7.5 (Runion, 2009). It was observed that the fresh cassava waste water kills plants in the farm. However when subjected to anaerobic digestion for a period of 30 days it can then be used in the farms as a good organic fertilizer for agriculture. The CW and SD (cassava waste water and swine dung blend) had a lower yield of 120L/total mass slurry; however it commenced flammability on the 10th day. The swine alone had a yield of 123 L/Total mass slurry and commenced flammability on the 6th day. The results showed that the animal waste had a positive effect on the cassava waste water since the CW on its own did not produce flammable gas. There was also attendant reduction in the foul odour of the waste after the digestion showing that the anaerobic digestion killed most of the pathogens responsible for the foul odour. Fig 3 shows the daily biogas production for the period, while Table 1 shows the lag period, cumulative and mean volume of gas productions. The lag period is the period from charging of the digester to onset of gas flammability (Ofoefule et al., 2010).

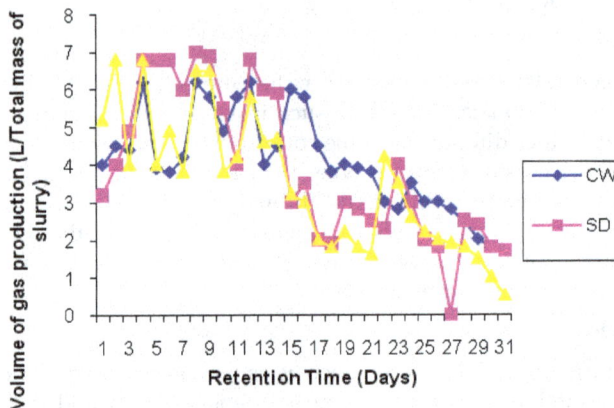

Fig. 3. Daily biogas production

PARAMETERS	CW	SD	CW : SD
Lag period (days).	Nil	5	9
Cumulative gas yield (L/ total mass of slurry).	130.25	122.55	119.90
Mean gas yield (L/ total mass of slurry).	4.20±1.32	3.95±2.01	3.87±1.80

Table 1. Lag period, Cumulative and mean volume of gas production of the pure wastes and blend

2.   **Effect of Abattoir cow liquor waste on biogas yield of some Agro-Industrial wastes.**
The results in this study showed that the cow liquor waste and cassava waste water blend (CLW: CW) did not flame throughout the retention period as a result of the acidic nature of the combined waste (pH=3.3). The carbonated soft drink sludge that commenced flammable biogas production on the 9th day stopped after one and half weeks as a result of the drop in pH from 5.68 to 5.20. The reduction in pH killed the microbes responsible for converting the

waste to biogas. However the CLW: BS (cow liquor waste: brewery spent grain) had the shortest onset of gas flammability and highest cumulative gas yield of 613.2 L/TMS (Table 2). Fig 4 shows the daily biogas production (Uzodinma and Ofoefule, 2008).

Parameters	BS	CS	CW	CLW : BS	CLW: CW	CLW: CS
Lag period (days)	20	8	Nil	6	9	8
Cumulative gas yield (L/TMS)	183.6	177.50	Nil	613.2	394.2	87.4
Mean Volume of gas yield (L/TMS)	7.34	7.10	Nil	24.53	8.23	2.54

Table 2. Lag periods, cumulative and mean volume of gas yield for single organic wastes and CLW blends

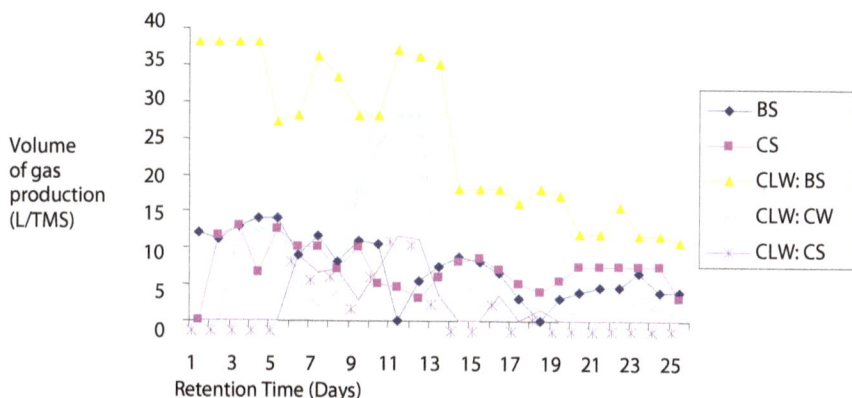

Fig. 4. Daily biogas yield

3. Preliminary studies on biogas production from blends of palm oil sludge with some Agro-based wastes.

The palm oil sludge (POS) in this study could not produce quantifiable gas within the 25 days retention period used for the experiment. However when combined with brewery spent grain (SG), carbonated soft drink sludge (SL) and cassava waste water (CW), reasonable quantities of biogas were produced which flamed after some lag periods as shown in Table 3. The POS: CW had the highest yield of biogas followed by the POS: SG while the least yield came from the POS:SL. The better yield of POS: CW over the others could be accounted for by the fact that the CW and others were allowed to be partially decomposed for a period of two months to increase their pH level, since in their fresh state they were found to be acidic. This resulted in the cassava waste water giving a better yield of biogas. Analysis of their flammable gas composition showed that POS: CW and POS: SL gave higher methane contents than POS: SG (Table 4). Fig. 5 shows the Daily biogas production (Uzodinma et al., 2007a).

Parameters	POS:SG	POSL:CW	POS:SL
Lag period (days)	10	8	15
Cumulative gas yield (L)	312	394.2	87.4
Mean volume of gas yield (L)	12.5	15.8	3.5

Table 3. lag periods, cumulative and mean volume of gas yield for POS blends

Waste blends	CO_2	CO	H_2S	CH_4
POS:SG	25.3	5.0	2.5	67.2
POS:CW	20.9	1.6	1.3	76.2
POS:SL	20.1	1.2	2.2	76.5

Table 4. Analysis of flammable gas composition for POS blends (%)

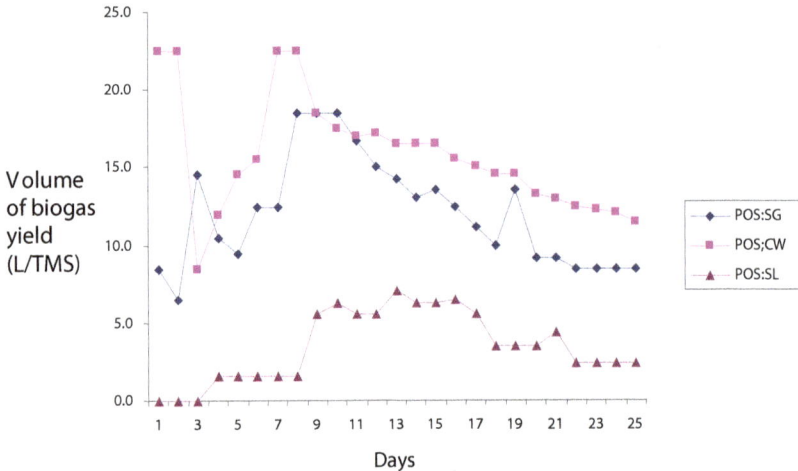

Fig. 5. Daily biogas yield for POS blends

4. Energy generation from microbial conversion of Treated cassava waste water from garri processing industry.

In this study, cassava waste water (CW) was treated with some other wastes to improve its pH level before digesting it. The waste used included; palm oil sludge (POS), powdered rice husk (RH) and pig dung (PD). The results showed that not only was the pH increased, the physicochemical properties also improved, which translated to higher biogas yields. The CW: RH gave the highest yield while the CW: PD followed with the shortest lag period of 4 days (Table 5). The higher yield of CW: RH was attributed to the fact that the rice husk was pre-decayed for about 1 month, and as a result had accumulated some microbes that aided in the faster digestion. The shortest lag period of CW: PD was explained by the fact that swine dung is a rumen animal, having the natural flora that are responsible for biogas production in its gut, aiding the fastest onset of gas flammability. Fig. 6 shows the Daily biogas production (Uzodinma et al., 2007b).

Parameters	CW:POS	CW:RH	CW:PD
Lag period (Days	8	6	4
Cumulative volume of gas Production (L/TMS)	394.20	481.30	432.00
Mean volume of gas production (L/TMS)	15.77	19.30	17.30

Table 5. Lag Period, Cumulative and Mean volume of biogas production

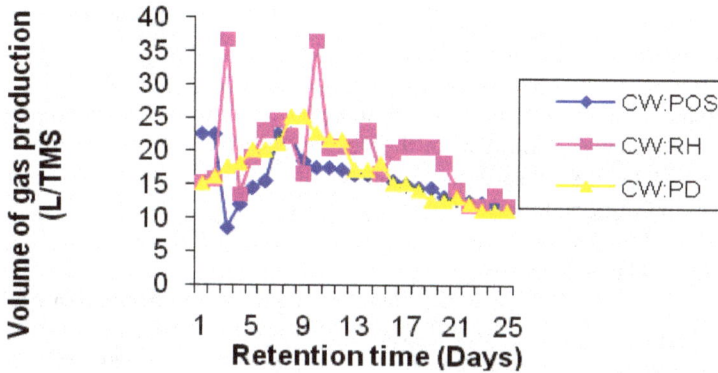

Fig. 6. Daily biogas production

Socio-Economic Benefits of Waste Water Treatment

Apart from reduction in environmental pollution from the treatment of waste waters, new demands for agricultural products arising from increased biomass usage would impact on the social-economic life of the populace especially when anaerobic digestion process of waste water treatment option is undertaken. Social issues such as employment generation, and poverty reduction especially for the developing countries would be addressed through this technology as a result of expanded economic activities across the real sector of the economy encompassing agriculture, manufacturing and exports. These would enhance people's ability to develop economic activities designated to reduce poverty particularly for the rural communities. Conversion of these biodegradable waste waters (both domestic and industrial) into biogas would result in cleaner air as well as efficient waste management system, improving the sanitary conditions of the urban environment. This will lead to socio-economic benefits with regard to health, income and security of the eco-system threatened by adverse climatic alterations (Ofoefule et al., 2009).

5. Conclusion

The results of these studies have shown that the waste waters/ slurries which are pollutants in the areas where they are processed can be sources of useful energy and organic fertilizers by subjecting them to anaerobic digestion for biogas production. The studies further revealed that most of these waste waters on their own are not capable of effective and efficient biogas production since they are mostly found to be acidic in their fresh states. They therefore need to be co-digested with other better producing wastes like animal wastes to enhance their flammable biogas production capabilities. The anaerobic digestion process of these waste waters is expected to be a source of waste management and pollution control.

6. References

Allison- Onyechere L.N., U. Ngodi and M.N. Ezike 2007. Anaerobic biotechnology for sustainable waste treatment. A review. J. Res. Bioscience. 3(1): 40-43.

Amuda, O.S and Ibrahim A.O .2006. Industrial waste water system using natural material as adsorbent.Africa Journal of Biotech Vol. (16), pp.1487-1487.

Anon, 1980. Survival of enteroviruses in rapid-infiltration basins during the land application of wastewater, Annl. Environ. Microbiol. 40:192-200, 1980.

Anon, 1995. "Environmental Estrogens: Consequences to Human Health and Wildlife". IEH assessment. Medical Research Council, Institute for Environment and Health.

Anon. 2005. "Environmental Hazards of the Textile Industry," Environmental Update #24, published by the Hazardous Substance Research Centers/ South & Southwest Outreach Program, June 2006; Business Week, June 5, 2005.

Anon. 2010. File:///D:/waste%20water/water%20 treatment%20method.htm, Buzzle.com, intelligent-life on web.

Avnimelech Y., Diab, S. and Kochba, M. 1993. Development and evaluation o a biofilter for turbid and nitrogen rich irrigation water. Wat. Res. 27: 785-790.

Bansode, R.R,Losso J.N, Marshall W.E, Rao, R.M and Portier R.J 2004. Pecan shell-based granulated activated carbon for treatment of chemical oxygen demand (COD), in municipal wastewater. Bioresource Technol. 94: 129-135.

Bhat, P.R., Chanakya, H.N. and Ravindranath, N.H., 2001. Biogas plant dissemination. J. Energy Sustainable Dev. 1: 39 – 41.

Chynoweth, D.P and Owens. M 1991. Biochemical methane potential of municipal solid waste components. Water Science Technology, 27:1-14.

Galambos II, Molina, J.M, Jaray, P., Vatai, G, Bekassy-Molner E. 2004. High organic content industrial wastewater treatment by membrane filtration. Desalination, 162:117-120. http://EzineArticles.com/?expert=Richard_Runion

Lettinga, G., Van Velsen, A.F.M., Hobme, W. de Zeeuw, J. and klapwijk, A. 1980. Use of the upflow sludge blanket (USB) reactor concept for biological waste water treatment especially for anaerobic treatment. Biotech and bioeng. 22:699-734.

Mara, D. and Horan, N., 2003. Water and Wastewater Microbiology, Academic Press.

Mehmet F.S and Hassan Z.S. 2002. Ozone treatment for textile effect and dyes: effect of applied ozone dose, pH and dye concentration. Journal of Chemical technology and Biotechnology, 77: 842-850.

Naik, A. 2010. Wastewater Treatment Methods. Accessed from http://www.buzzle.com/chapters/chapters.asp. 9th of July 2010.

Namboodri C.G, Perkins W.S and Walsh, W.K. 1994. Decolonizing dyes with chlorine and ozone: Part II, American dyestuff Report, 83: 17-26.

Nasim, A.K, Shaliza,I and Piarapakaran,S. 2004. Elimination of heavy metals from wastewater using agricultural wastes as adsorbents. Malaysian Journal of Science, 23:43-51.

Nomanbhay, S.M, Palanisamy, K. 2005. Removal of heavy metals from industrial wastewater using chitosen coated oil palm shell charcoal. Electronic Journal of Biotechnology, 8:43-53.

Oboh, G. 2005. Isolation and characterization of amylase from fermented cassava (Manihot esculenta Crantz) waste water. African journal of biotechnology. Vol 4 (10), pp 1117-1123. ISSN 1684-5315.

Ofoefule, A.U., Chima, P.U., Nnabuchi, M.N. and Uzodinma, E.O. 2010. Anaerobic batch co-digestion of cassava waste water and Swine dung. Nig. J. Solar Energy 20:128 – 132

Ofoefule, A.U., Ibeto C.N., Uzoma, C.C., Oparaku, O.U. 2009. Biomass Technology: A key driver for improving climate change and socio-economic life in Nigeria. Int. J. Environ. Sci. 5 (1): 54- 58.

Okafor N. 1998. An integrated bio-system for the disposal of cassava wastes. Proceedings of the internet conference on integrated bio-systems, edited by Eng-Leong, F and Tarcisio, D.S. Pp.1-5.

Okieimen, F.E, Ogbeifun, D.E, Navala,G.N, Kumash, C.A. 1985. Binding of copper, Cadmum and Lead by modified cellulosic materials. Bull., Environ. Contam. Toxicol.34: 860-870.

Oparaku, N.F, Mgbenka, B.O and Ibeto, C.N. 2011. Wastewater disinfection utilizing ultraviolet light. Journal of Environmental Science and Technology,4(1):73-78. DOI:10.3923/jest.2011.73.78.

Peres, J.A, Beltran de Heredia, J, Dominguez, J.R. 2004. Integrated Fenton's reagent-coagulation/flocculation process for treatment of cork processing wastewaters. J. .Haz. Mat. 107 (3):115-121.

Preetha, B., Viruthagiri, T. 2005. Biosorption of Zinc (II) By *Rhizopus* equilibrium and kinetic modeling. African J.Biotechnol. 4(6): 506-508.

Rajeswari, K.R. 2000. Ozonation treatment of textile dyes wastewater using plasma ozonizer, PhD thesis, University of Malaysia, Malaysia.

Runion R. 2009. All about Waste water treatment.

Runion R. 2010. Wastewater - Contamination Sources. http://www-all-about-wastewater-treatment.com/category/wastewater.

Stanislaw L. and Monica G. 1999. Optimization of oxidant dose for combined chemical and biological treatment of textile waste water. Water Research 33, 2511-2516.

Tam,M and Antal M. 1999. Preparation of activated carbons from macademia nut shell and coconut shell by air activation. Ind.Eng.Chem Research, 38:4268-4276.

Uzodinma E.O. and Ofoefule, A.U. 2008. Effect of abattoir cow liquor waste on biogas yield of some Agro-industrial wastes. Sci. Res & Essay 3 (10): 473-476.

Uzodinma E.O. Eze, J.I. and Onwuka, N.D. 2007b. Energy generation from microbial conversion of treated cassava (manifot utilissima) waste water from garri processing industry. J. Res in Bio Sci 3(1):61-66.

Uzodinma E.O., Ofoefule, A.U., Eze, J.I. and Onwuka, N.D. 2007a. Preliminary studies from blends of palm oil sludge with some Agro-based wastes. Nig. J. Solar. Energy 18: 116-120.

Van Haandel, A.C. and Lettinga, G. 1994. Anaerobic sewage treatment. A practical guide for regions with a hot climate. John Wiley and Sons, Ltd. Chichester ISBN 0-471-95121-8.

Verstraete, W. deBeer, D., Pena, M., Lettinga, G. and ens, P. 1996. Anaerobic bioprocessing of organic waste. Accepted for Publication in World J. Microbio and Biotech.

Verstraete, W. and Top, E. 1992. Holistic environmental technology. In: Microbial control of pollution. CJ. Fry, G. Gadd, R. Herbert and I. Watson-Crack eds), Cambridge Univ. Press.pp 1-18.

Wolfgang, M. and Axel, H. 2005. An introduction to anaerobic digestion. Seminar presented at the Biowaste. Digesting the alternative seminar UK.

6

Agricultural Dairy Wastewaters

Owen Fenton, Mark G. Healy, Raymond B. Brennan, Ana Joao Serrenho,
Stan T.J. Lalor, Daire O hUallacháin and Karl G. Richards
Teagasc, Environmental Research Centre, Wexford
National University of Ireland, Galway
Rep. of Ireland

1. Introduction

In Ireland, farming is an important national industry that involves approximately 270,000 people, 6.191 million cattle, 4.257 million sheep, 1.678 million pigs and 10.7 million poultry (CSO, 2006). Agriculture utilizes 64% of Ireland's land area (Fingleton and Cushion, 1999), of which 91% is devoted to grass, silage and hay, and rough grazing (DAFF, 2003). Grass-based rearing of cattle and sheep dominates the industry (EPA, 2004). Livestock production is associated with external inputs of nutrients. Phosphorus (P) surpluses accumulate in the soil (Culleton et al., 2000) and contribute to P loss to surface and groundwater (Tunney, 1990; Regan et al., 2010). Elevated soil P status has been identified as one of the dominant P pressures in Ireland (Tunney et al., 2000). Schulte et al. (2010) showed that it may take many years for elevated soil P concentrations to be reduced to agronomically and environmentally optimum levels. The extent of these delays was predominantly related to the relative annual P-balance (P balance relative to total P reserves). While the onset of reductions in excessive soil P levels may be observed within five years, this reduction is a slow process and may take years to decades to be completed.

Agricultural wastes and in particular dairy slurry and dirty water are discussed in this chapter. However, while the term 'waste' is commonly used for these materials, it is an unfortunate label, as it suggests that the materials have no further use and are merely a nuisance by-product of farming systems that must be managed. However, given the high nutrient contents of these materials, it is far more appropriate for them to be considered as organic fertilizers, and as such being a valuable commodity for the farmer. With higher and more volatile chemical fertilizer prices in recent years, the fertilizer replacement value in economic terms of these materials is increasing. Therefore, the management of agricultural 'wastes' in a manner that maximises the nutrient recovery and fertilizer value to crops should be a priority within any management plan for these materials.

Nutrient contents and various research areas regarding management, remediation and control of such nutrients to prevent losses to the environment are discussed. The Surface Water Directive, 75/440/EEC (EEC, 1975), the Groundwater Directive, 80/68/EEC (EEC, 1980), the Drinking Water Directive, 98/83/EC (EC, 1998), the Nitrates Directive, 91/676/EEC (EEC, 1991(a)) and the Urban Wastewater Directive, 91/271/EEC (EEC, 1991(b)), combined with recent proceedings taken against the Irish State by the EU Commission alleging non-implementation of some aspects of the directives, has focused considerable attention on the environmentally-safe disposal of agricultural wastewaters in Ireland. To address these directives, the WFD (2000/60/EC, 2000) came into force on 22nd December, 2000 and was transposed into Irish legislation by the European Communities (Water Policy) Regulations 2003 on the 22nd December, 2003. Eight "River Basin Districts" (RBD) were established in Ireland, north and south, with the aim of achieving "good status" in all surface and groundwater by 2015. The WFD will bring about major changes in the regulation and management of Europe's water resources. Major changes include:

- A requirement for the preparation of integrated catchment management plans, with remits extending over point and non-point pollution, water abstraction and land use;
- The introduction of an EU-wide target of "good ecological status" for all surface and groundwater, except where exemptions for "heavily-modified" water bodies are granted. Programmes of measures (POM) must be put in place to protect groundwater and surface water while being efficient and cost-effective. POM to achieve at least "good ecological status" must be implemented by the agricultural sector by 2012. In Ireland the Nitrates Directive is the main POM in place. At present, a strategy exists within Europe to restore the "good ecological status" of surface and groundwater. It focuses on reducing nutrient pressures to prevent further nutrient loss to surface and groundwater. However, intensification of agriculture poses a challenge to the sustainable management of soils, water resources, and biodiversity. N losses from agricultural areas can contribute to ground- and surface water pollution (Stark and Richards, 2008; Humphreys et al., 2008).

Results from a Water4all project suggest that regulation alone will not achieve sufficient reduction in water quality as nitrate builds up in soils and the long residence time of groundwater in aquifers needs a more immediate solution (Water4all, 2005; Hiscock et al., 2007). Therefore, remediation (nitrogen - N) and control (phosphorus – P) technologies must be an integral part of the process for point and diffuse pollution from historic or future incidental nutrient losses. Solutions developed must be integrated efforts within a catchment or river basin.

Good Agricultural Practice Regulations under The Nitrates Directive (European Council, 1991) is currently the main mitigation measure in place within the agricultural sector to achieve the goals of the WFD. These regulations came into effect in the Republic of Ireland in 2006 under Statutory Instrument (S.I) 788 of 2005, and subsequently under S.I 378 of 2006, S.I 101 of 2009 and S.I 610 of 2010. The Nitrates Directive sets limits on stocking rates on farms in terms of the quantity of N from livestock manure that can be applied mechanically or directly deposited by grazing livestock on agricultural land. A limit of 170 kg N ha^{-1} year^{-1} from livestock manure was set. However, the EU Nitrates Committee approved Ireland's application for a derogation of this limit to allow grassland-based (mostly dairy) farmers to operate at up to 250 kg N ha^{-1} year^{-1} from livestock manures, with the understanding that this derogation will not impinge on meeting the requirements of the Nitrates Directive. The current average stocking density on dairy farms is 1.81 livestock units (LU) ha^{-1}.

The "Good Agricultural Practice for the Protection of Waters" regulation, S.I 778 of 2005 (Anon, 2005), came into effect on February 1st 2006. The most recent revision of the regulation was published in 2010 (Anon, 2010). It constrains the use of P and N fertilizers, ploughing periods and supports derogation on livestock intensity. In particular it regulates farmyard and nutrient management, but also examines prevention of water pollution from fertilizers and certain activities. The linkage between source and pathway can be broken if pollutants remain within farm boundaries and are not discharging to drainage channels, subsurface drainage systems, or entering streams or open waterways within farm boundaries. These regulations also place restrictions on land spreading of agricultural wastes. This strategy looks at present loss and future loss prevention. There are no guidelines in place for the remediation or control of contaminated discharges to surface and/or groundwater or future discharges due to incidental losses. Traditionally, agricultural wastes are managed by land spreading. Following land spreading, the recharge rate, the time of year of application, the hydraulic conductivity of the soil, the depth of soil to the water table and/or bedrock, and the concentration of nutrients and suspended sediment in the wastewater (dirty water and any discharge containing nutrients) are some of the defining parameters that determine nitrate movement through the soil to the watertable. The maximum instantaneous rate of application is 5 mm per hour and the quantity applied should not exceed 50 m³ per hectare per application (ADAS, 1985; 1994; DAFF, 1996) and these recommendations are present within best farm management practices. Infiltration depth of irrigated water and rainfall may be estimated when the annual effective drainage, number of effective drainage days, effective porosity, annual precipitation, and the hydraulic load of the irrigator are known (Fenton et al., 2009(b)). This data may then be combined with watertable data to examine if excess nutrients recharge to groundwater within a specific time frame.

2. Agricultural dairy wastes

2.1 Types of dairy wastes and nutrient content

In a grassland system, the N recovery rate of dairy slurry is highly variable due to variations in slurry composition, application methods, spreading rates, soil and climatic conditions and slurry N mineralisation rates (Schröder, 2005). In Ireland, approximately 80% of manures produced in winter are managed as slurries containing 70 g kg⁻¹ dry matter, 3.6 g kg⁻¹ total N (TN) and 0.6 g kg⁻¹ total P (TP) (Lalor et al., 2010(a)). About 50% of the TN is in ammoniacal form and has the potential to be volatilised as ammonia during storage and following land spreading. Estimated organic managed waste generation for Ireland is presented in Table 1.

Waste Category	Waste Generation	
	Tonnes wet weight	%
Cattle manure and slurry	36,443,603	60.6
Water (dairy only)	18,377,550	30.5
Pig slurry	2,431,819	4.0
Silage effluent	1,139,231	1.9
Poultry litter	172,435	0.3
Sheep manure	1,336,336	2.2
Spent mushroom compost	274,050	0.5
Total	60,170,025	

Table 1. Estimated agricultural organic managed waste generation in 2001 (EPA, 2004a).

Great variation in the nutrient content of dairy slurry exists depending on feed type, age of sample when tested, age of the animal and how the effluent is stored and managed (Smith and Chambers, 1993). Seasonal differences in nutrient contents also exist (Demanet et al., 1999). Tables of published slurry nutrient contents in Europe exist (see MAFF, 2000). Such values are similar to South American dairy slurry concentrations found by Salazar et al. (2007). Some dairy slurry concentrations for undigested and digested samples are presented in Table 2. These tend to be similar to other nutrient contents across Europe found by Villar et al. (1979); Scotford et al. (1998(ab)) and Provolo and Martínez-Suller (2007). In Ireland, dirty water is generated from dairy parlour water and machine washings, precipitation and water from concreted holding yards (Photo 1). Average dirty water production per cow is 49 L^{-1} day^{-1}. Although dilute, dirty water has sufficient nutrients to give rise to eutrophication if lost to a waterbody through runoff or excess infiltration. Implementation of current legislation requires separation of faecal matter and water, thus diminishing the nutrient content of dirty water for land application (Photo 1). As the nutrient content is reduced and storage and water charges are high, an alternative solution to dirty water management is remediation and re-use for washing yards (Fenton et al., 2009). A number of papers have reported the chemical composition of dirty water from dairy farms (ADAS, 1994; Cumby, 1999; Ryan, 2006; Fenton et al., 2009(a);Minogue et al., 2010). Table 3 presents a range of nutrient contents available in dirty water from a number of studies. Minogue et al. (2010) and Cumby (1999) report higher mean TN nutrient figures for 20 farms in England and Wales of 580±487 mg TN L^{-1}. Martínez-Suller et al. (2010(b)) reviewed the composition of dirty water in the literature including others not mentioned in Table 3.

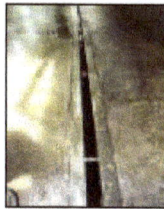

Photo 1. Dirty water generation: wash down high volume low pressure hose and drainage channel for speeding up washing after milking (Source: www.teagasc.ie)

Prediction of the nutrient content of agricultural waste waters would help farmers to more accurately calculate the nutrient fertiliser replacement value of the landspread materials and the additional fertiliser requirements for their crops. Martínez-Suller et al. (2010(a)) suggest that dry matter content or electrical conductivity are rapid, cheap methods to estimate the nutrient content of waste waters and manures.

2.2 Faecal microorganisms

Agricultural wastes not only pose a threat to waterbodies, a second major concern is the presence of pathogenic and/or antibiotic resistant bacteria in animal wastes (Sapkota et al., 2007) and the threat to human health. If properly handled and treated, manure is an effective and safe fertiliser. However, if untreated or improperly treated, manure may become a source of pathogens that may contaminate soil, food-stuffs, and water bodies (Vanotti et al., 2007). Animal manures are known to contain pathogenic bacteria, viruses and parasites (Pell, 1997). The contamination of surface waters with pathogenic micro-organisms transported from fields to which livestock slurries and manure have been applied is a serious environmental concern as it may lead to humans being exposed to such micro-organisms via drinking water (Skerrett and Holland, 2000); bathing waters (Baudart et al., 2000); and water used for the irrigation of ready to eat foods (Tyrel, 1999). A recent study

Table 2. Dairy slurry nutrient content from various studies.

Study	Manure Type	Digested/ Undigested	NH4-N	Organic N	NO3-N	TN	TP
			mg L⁻¹	mg L⁻¹	mg L⁻¹	mg L⁻¹	mg L⁻¹
Mulbry and Wilkie, 2001	Dairy	Undigested	306	904	<1	1210	303
		Digested (Maryland, USA)	1620	751	<1	2371	240
		Digested (Florida, USA)	178	47	<1	225	24.7
Pizarro et al., 2002	Dairy	Digested	5 to 80	-	-	-	1 to 20
Wang et al., 2009	Dairy	Digested (Minnesota)	2232			3456	249.7
	Dairy	Undigested (Minnesota)	1782			3305	266
Kebede-Westhead et al., 2004	Dairy	Digested manure effluent (Florida, USA)	233		<1	412	64.5
Brennan et al., 2010(abc)	Dairy	Undigested (West Ireland) May	1759(123)			4433 (271)	1138 (76)
	Dairy	Undigested (West Ireland) July				4861(425)	1144 (93)
	Dairy	Undigested (West Ireland) December				3982(274)	811(37)
Martinez -Suller et al., 2010	Dairy	Undigested (South East Ireland)				3430(1400)	560(250)

() standard deviation

Table 3. Dairy dirty water nutrient results from various studies in U.K. and Ireland. Based on new data and Martinez-Suller et al., 2010(b). All units in mg L^{-1}

Study	Number of farms	Period of study		BOD$_5$	K	NH$_4$-N	P	TP	TN
Minogue et al., 2010	60-Ireland	Monthly for 1 year	Mean	2246	568	212	37	80	587
			Min	0	3	0	0	2	0
			Max	19085	7232	2933	1240	795	6030
			SD	2112	513	206	53	68	536
Martinez-Suller et al., 2010(b)	1-Ireland	13 weeks (January – May)	Mean	3084	415	32	8.5	44	351
			Min	1570	213	0	0.7	21	128
			Max	8400	977	106	24.9	103	987
			SD	1739	169	25	6.49	22	231
Serrenho et al., 2010	1-Ireland								
Fenton et al., 2009(a)	1-Ireland	3 months (August - October)	Mean	-	-	-	14.0(9.2)	20.1 (6.9)	170(33.2)
Singh et al., 2005	1-Ireland			3000	-	-		111	479
Dunne et al., 2005(ab)	1-Ireland	Winter		2828		42			
		Spring		2703		53			
		Summer		2682		36			
		Autumn		2303		6			
Ryan et al., 2005	1-Ireland		Min	-	-	-	-	-	43
			Max	-	-	-	-	-	126
Rodgers et al., 2003	1-Ireland			2208					
Cannon et al., 2000				1440					
Cumby et al., 1999	20- England and Wales	Summer		9670	150	58	-	49	95
		Autumn		7450	85	48	-	34	70
Richards, 1999	1-Ireland					84.4			188
Misselbrook et al., 1995	1-England				350	42		<100	450
Ryan, 1991	1-Ireland			2077	210	92	17	23	159

(Venglovsky et al., 2009) has shown that animal manure contributes significantly to pathogen loading of soil and consequently runoff to waterways. Furthermore, a recent report by the EPA in Ireland (Lucey, 2009) highlighted land-spreading of manure or slurry as one of the main sources of microbial pathogens in groundwater. Additionally, a report by the Food Safety Authority of Ireland (FSAI, 2008) stated that 'there is potential for the transfer of pathogens to food and water as a result of land-spreading of organic agricultural material'.

Research from New Zealand, shows that dirty water contains faecal micro-organisms, which originate from dairy cattle excreta. Researchers such as Aislabie et al. (2001); McLeod et al. (2003) and Donnison and Ross (2003) have shown transfer of bacterial indicators, faecal coliforms and *Campylobacter jejuni* through soil. The Pathogen Transmission Routes Research Programme in New Zealand showed that significant faeces contamination arose through the deposition of faeces by grazing animals with access to waterways. Fencing and implementation of buffer strips were recommended as mitigation measures to prevent such losses (Collins et al., 2007). Presence of faecal indicator organisms is used to identify waters impacted by faecal matter from mammals. Indicators of faecal contamination such as *E. coli* are widely used as they are faecally specific and believed to not survive for more than 4 months post excretion (Jamieson et al., 2002). Recent research has shown that *E. coli* can survive for long periods of time in temperate soils (Brennan et al., 2010) and contribute to high detections in drainage waters from agricultural soils. *E. coli* were particularly associated with poorly drained soils due to the greater persistence of preferential flow channels and anaerobic micro-sites where they might survive. Thus the presence of *E. coli* in waters may not indicate recent contamination by faecal matter but could be due to historical pathogen deposition. Many treatment systems may be used to treat livestock waste and remove or decrease viral, bacterial and eukaryotic pathogens. Examples include bio-gas producing anaerobic digestion, composting, aeration, storage under a variety of redox conditions, and anoxic lagoons, all of which have been reviewed by Topp et al. (2009).

2.3 Current management practices for agricultural waste waters

The Nitrates Directive and rising costs are now forcing better use of nutrients in slurry. Research in the U.K. (Misselbrook et al., 1996; 2002; Smith and Chambers, 1993; Smith et al., 2000) includes improving N recovery from slurry by examining the effect of spreading method and timing, and reducing ammonia (NH_3) losses from slurry by evaluating splash-plate versus alternative techniques such as trailing shoe or trailing hose slurry application methods. The average abatement of these methods varies and differs when grassland or arable application are considered (Smith and Misselbrook, 2000; Misselbrook et al., 2002). Present research in Ireland follows similar patterns (Ryan, 2005). Ammonia emissions with respect to trailing shoe versus splash-plate and subsequent N uptake by the sward are being investigated in Irish grasslands (Lalor and Schulte, 2008). Farm management strategies aimed at prevention of nutrient loss to water have recently been reviewed by Schulte (2006). The Nitrates Directive regulations impose limits to N and P inputs onto livestock and tillage farms. Cattle and dairy farming systems are required to make more efficient use of nutrients. International experience suggests that significant gains in nutrient efficiency can be made by increasing the utilisation of N in slurry. Lalor (2010(b)) suggested N-utilisation efficiencies from slurry as low as 5% under existing practices, whereas international literature suggests that there is scope to raise efficiencies to 40-80%. Despite the relatively low utilisation in practice, the Nitrates regulations set a nitrogen fertilizer replacement value

(NFRV) target of 40%, presenting a considerable challenge for the grassland sector. In addition, the ceiling to nutrient inputs imposed under the Nitrates Directives made it difficult for many livestock farmers to continue to accept pig slurry as a fertilizer onto their farm. In Ireland, as a result, the potential for the traditional practice of spreading slurry on grasslands has been reduced significantly. Returning pig slurry to arable land allows a more closed nutrient cycle to operate, since cereal grains constitute a significant proportion of the diet of pigs. However, this creates a major logistic challenge where arable land and pig farms are not closely located (Lalor et al., 2010(b)).

In an Irish study, cattle slurry application on grassland shows that the NFRV in the year of application is affected by application method and timing. Cattle slurry applied (using traditional methods) with splashplate had an NFRV of 21% in April and 12 % in June. Application using trailing shoe (a modern alternative which places slurry in thin bands along the soil surface) increased the NFRV to 30% in April and 22% in June. Changing application timing from summer to spring with existing splashplate machinery is the most cost effective strategy for improving NFRV. Approximately 4% of the total slurry N applied was recovered in the second year after application. For repeated applications over a number of years, models indicate that the maximum cumulative residual recovery would be 12-14% of the annual slurry N application rate. It would take approximately 10 years of repeated slurry applications for the residual N release to reach this maximum level (Lalor et al., 2010(b)). In Ireland, research by Lalor et al. (2010(b)) showed that the NFRV target of 40% set in the Nitrates regulations can only be achieved when the residual N release is included, and when best practice strategy of trailing shoe application in April was adopted. Spring application of slurry is often restricted by soil trafficability, particularly on poorly drained soils. The trailing shoe application method can provide more flexibility for spring application as grass contamination is reduced compared to splashplate.

In Ireland, besides land application methods (splashplate or trailing shoe (Photo 3)), dirty water irrigation using centre pivotal irrigation systems is common (Photo 2). The recommended irrigation rates should not exceed 5 mm hr^{-1}. Strict guidelines for their safe use are in place. Application timing of dirty water should take soil moisture status and soil physical properties into account (Houlbrooke et al., 2004). Two pond systems are used in many countries reducing the biological oxygen demand and suspended solids contents. A limitation here is that the nutrients remain unchanged and need to be landspread with potential environmental consequences. An upgraded "advanced pond system" has been

Photo 2. Rotational centre pivot sprinkler system used for dairy dirty water irrigation (Source: www.teagasc.ie)

Photo 3. Slurry tanker with trailing shoe application system (Source: Teagasc)

designed by Craggs et al. (2004) and could be an alternative on dairy farms. Houlbrooke et al. (2006) showed that individual irrigation systems with low intermittent irrigation rates (0.4 to 4 mm hr^{-1}) could be used without nutrient losses. To facilitate this low irrigation rate, increased storage is needed on a farm. Adapted low irrigation lines have now also been investigated, the position of which may be changed through use of a quad-bike system.

Bolan and Swain (2004) reviewed issues and innovations in land application of farm wastes in New Zealand and showed that research must focus on improved systems to convert manure based wastes into a valuable but also environmentally benign product.

An alternative manure management system in some countries is anaerobic digestion. Manures are an excellent source of organic material for anaerobic digestion and the production of bio-gas. Co-digestion of agricultural wastes with sewage sludges can further improve the methane production in anaerobic digesters (Ward et al., 2008).

2.4 Environmental Impact of agricultural waste waters

Agricultural waste waters can contain N, P, K, S, C, pathogenic micro-organisms and a range of other micro nutrients. Nutrients returned to agricultural soils through land-spreading are important for nutrient efficiency on farms and for reducing reliance on inorganic fertilisers. Land application should be at rates that supply nutrients for crop growth and at the time when these nutrients are required. Addition of excessive nutrients at times of reduced crop demand can increase the potential for losses of nutrients such as N and P, which contribute to surface water eutrophication and can lead to pollution of drinking waters. In addition, land application to wet soils can lead to increased emissions of greenhouse gases such as nitrous oxide (N_2O).

Landspreading of dairy slurry and wastewaters has been associated with ammonia volatilisation to the atmosphere. Application of ammoniacal nitrogen (NH_4^+) to soils in wastewater increases the soil solution NH_4^+ concentrations, which is in equilibrium with free ammonia (NH_3) which is also in equilibrium with the concentrations in the atmosphere (See Equation 1).

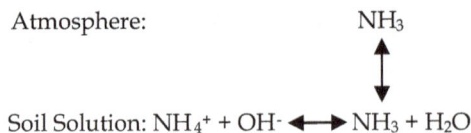

$$\text{Atmosphere:} \qquad\qquad NH_3$$

$$\updownarrow$$

$$\text{Soil Solution: } NH_4^+ + OH^- \longleftrightarrow NH_3 + H_2O$$

Ammonia volatilisation from soil lowers the pH of soil directly under the waste water. Further soil pH reduction can also occur when the volatilised NH_3 is re-deposited and nitrified. Agriculture is the main emitter of NH_3 to the atmosphere accounting for ~80% of total global emissions (Stark and Richards, 2008) and is expected to reach 109 Tg N yr^{-1} by 2050. Once in the atmosphere, NH_3 can readily combine with NO_3 and SO_4^{2-} in acid cloud droplets to form particulates and can be transported over long distances before being deposited again to soil or water. Atmospheric N deposition has increased over recent decades and ranges from 5 to 80 kg ha^{-1} yr^{-1} with a global average of 17 kg ha^{-1} yr^{-1} have been observed. Deposited NH_3 can then lead to acidification of soil and eutrophication of waters, which led to the UN establishing the Convention on Long-range Transboundary Air Pollution including NH_3 and the EU set limits for NH_3 from European countries. Emissions of NH_3 from agricultural slurry and waste waters can be reduced through utilisation of low emission storage facilities where stores are covered to reduce contact with the atmosphere. Emissions can also be reduced from the field through the use of low emissions spreading methods such as band spreading and injection. Land application of dilute waste wasters has lower NH_3 emissions compared to more solid waste due to a reduction in the NH_4^+ content and the infiltration of the liquid waste into soil, reducing atmospheric contact. Thus dilute effluents have lower NH_3 emissions, but potentially greater N_2O and NO_3^- emissions.

Application of animal slurries and wastewaters to soils promotes denitrification through the supply of readily available N and C for microbial respiration and also by promoting anaerobic conditions in the soil through partial sealing of soil pores and the consumption of oxygen through C oxidation. Storage of manures leads to the build up of volatile fatty acids which are readily degradable forms of C. Microbial denitrification associated with landspreading of organic wastes can be an important source of the potent greenhouse gas, N_2O. Emissions of N_2O from slurry spreading are mainly related to the application method, and the soil temperature combined with soil moisture at the time of application. Methods for reducing N_2O emissions associated with waste waters include limiting the hydraulic loading to ensure soils remain aerobic; adjusting application timing to when soils are not anaerobic; adjusting application method/rate; inclusion of nitrification inhibitors to slow the rate of NO_3^- formation; manipulation of the C/N ratio; digestion or storage to reduce labile C content and inclusion of materials with high cation exchange capacity e.g. zeolite. A schematic of soil N transformations is presented in Figure 1.

There have been numerous reports of water pollution occurring after landspreading of wastewaters to soils. Richards et al. (2004) reported nitrate leaching losses ranging from 95 to 323 kg N ha^{-1} when wastewaters were over-applied to free draining soils. Houlbrouke et al (2004) reported between 2 and 20% of N and P applied in agricultural wastewaters leached through soils and the concentrations leaching were above ecological limits for good water quality.

Repeated application of wastewaters to soil can lead to an increase in the organic fractions of N, P, K and organic carbon due to changes in soil organic matter. In New Zealand, Barkle et al. (2000) reported significant increases in soil total N and organic C. At low temperatures increasing soil C content due to dirty water application can lead to greater N immobilisation due to changes in the soil C/N ratio (Ghani et al., 2005). Increasing soil nutrient status above the agronomic optimum has been shown to increase the risk of nutrient loss to water (Sharpley and Tunney, 2000). Other soil properties can be influenced by land application such as increasing soil pH, changes in soil hydraulic conductivity due to clogging, plugging and macropore/aggregate collapse. Often the actual effect of landspreading on soil physical properties is difficult to quantify due to variability in soil physical properties, short term observation and experimental approaches within a background of seasonal variation in properties (Hawke and Summers, 2006) Agricultural waste waters can contain high concentrations of pathogen micro-organisms such as Campylobacter, Listeria, Cryptosporidium and Salmonella spp. Loss of high concentrations of faecal pathogens to water can result in the waters being unfit for human consumption and failing to meet water quality standards for bathing water quality. Pathogen transfer to water can occur when waste waters are applied to water-logged soils where water flow over soil leads to high pathogen losses to rivers and associated bathing waters (Kay et al., 2007). Reducing the volume and area contaminated by waste waters on farms can reduce emission of pathogens to water by 10% (Kay et al., 2007).

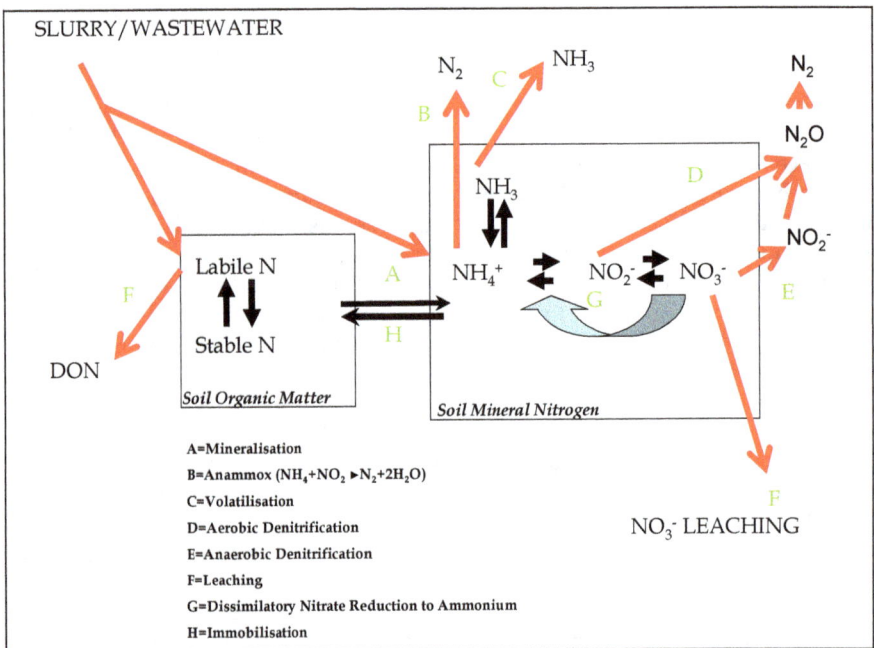

Fig. 1. Soil N transformations of slurry/wastewater derived nitrogen inputs.

3. Novel remediation techniques currently being researched

Fenton et al. (2008) reviewed agricultural wastewater remediation and control technologies suitable for Ireland. Several options such as use of chemical amendments, subsurface carbon emplacement and wetlands were some of the options proposed for further research.

3.1 Amendments to dairy slurry and dirty water

Dairy dirty water is a bio-product of dairy farming. The usual method for disposal of this product is land-spreading (Healy et al., 2007). This can increase the P concentration on the soil surface and the pollution related with the natural run-off during rain events. Not many studies have been made regarding this subject.

Due to the properties of the dairy dirty water, the potential for leaching should also be considered. Usually, P leaching is not considered to be a significant problem in groundwater because it is not very mobile in soils or sediments, and should therefore be retained in the soil zone. However, in extremely vulnerable areas, where the soil and subsoil are shallow and where P enters groundwater in significant quantities, groundwater may act as an additional nutrient enrichment pathway for receptors such as lakes, rivers and wetlands (EPA, 2008). Phosphorus leaching may occur in sandy soils (Carlyle et al., 1998) or where there are preferential flow paths in the soil.

In the past, the primary objective of chemical amendment of manure was to reduce NH_3 losses from manure as this increased N availability to plants. In recent years environmental concerns have shifted this focus to amendments, which mitigate P loss from soils and manure. In Ireland, the focus of recent research has been to find amendments which reduce solubility of P in dairy cattle slurry in particular. The use of such amendments must be practical and cost effective for the farmer. The effect of reducing P solubility on reducing subsequent P fertilizer replacement value of the material should also be considered.

Alum (aluminium sulphate) has been used extensively to treat poultry litter in the U.S for over 30 years with great success to reduce NH_3 in poultry houses and reduce soluble P in poultry litter (Moore and Edwards, 2007). These authors also found that alum addition to poultry litter reduced P loss, ammonia volatilisation and had negligible effect on metal release from amended soil. Work involving amendments of swine and dairy cattle slurries for the control of P have been limited to laboratory batch studies with little emphasis on cost or feasibility of treatments (Dao, 1999; Dou et al., 2003; Kalbasi and Karthikeyan, 2004; Smith et al., 2001; Moore et al., 1998).

Aluminium chloride has been recommended as the most suitable amendment for controlling P solubility in swine and cattle slurry (Smith et al., 2001). In an incubation study Dou et al. (2003) found that technical grade alum added at 0.1 kg/kg (kg alum per kg slurry) and 0.25 kg/kg reduced Water Extractible P in dairy and swine slurry by 99% and 80%, respectively. Dao (1999) amended farm yard manure with calcium carbonate, alum and fly ash in an incubation experiment and reported WEP reductions in amended manure compared to the control of 21, 60 and 85%. Penn et al. (2009) examined the sorption and retention mechanisms of several P sorbing materials (PSMs) including acid mine drainage treatment residuals, water treatment residuals, fly ash, bauxite mining residual and FGD in lab experiments and found the degree of sorption of P to be strongly influenced by the solution pH, buffer capacity of manure, and ionic strength of amendments. These amendments are attractive as they are free. However, they are more variable than chemicals and commercial coagulants used by other workers and much more research is required before there could be used in practice. Internationally, P sorbing amendments have been used to control P losses after manure application. P sorbing amendments can either be added directly to the manure before land application (Moore et al., 1998), spread on the ground before manure application (McFarland et al., 2003), or incorporated into the topsoil at (Novak and Watts, 2005).

Ochre from coal mining origins in the U.K. is a low value waste product from acid mine drainage and has been used as an amendment to sequester P in filters or drainage ditches, or in wetlands receiving sewage or agricultural waste. In Ireland, metal release from metal mining Avoca ochre has made it unsuitable for environmental purpose (Fenton et al., 2008; Fenton et al., 2009(a)). Ochre has a high P sequestration capacity with 97% of sequestration occurring within 5 minutes of contact with an agricultural waste.

3.2 PRB and reactive media for enhanced denitrification

Low-cost, in situ treatment systems, called permeable reactive barriers (PRB), may be used to treat groundwater. In these systems, N-rich wastewater flows through a carbon (C)-rich mixture to reduce nitrate concentrations to acceptable levels. Organic C amendments offer low-cost surface and subsurface treatment alternatives for wastewater treatment. C availability is an important factor that affects denitrifying activity in soils. The presence of C provides an energy source, thereby enhancing the potential for denitrification. Denitrification may be increased in soils by the addition of an external C amendment. This amendment may be natural C such as woodchip, wheat straw, corn, vegetable oil, sawdust mulch, or other materials, such as treated newspaper or unprocessed cotton (Volokita, 1996). A PRB or denitrification wall is only one of many denitrifying bioreactor types, i.e. denitrification beds, up-flow bioreactors, stream bed bioreactor or denitrification layers. The limitations of a denitrification wall are that they require site specific analyses of hydraulic gradient, the depth and extent of the nitrate plume/s, removal of nitrate is confined to up-gradient pollution sources and within the upper 2 m of groundwater. Problems may arise if the denitrification wall has a lower saturated hydraulic conductivity than the surrounding sub-soil. If this occurs, nitrate plumes tend to flow around the wall and not through it. However, in cases where nitrate contamination occurs below 2 m, the diameter (parallel to flow path of contaminant) of the trench may be widened. This causes up-welling into the more permeable trench. Flow through these denitrification bioreactor systems may be either horizontal or vertical. In laboratory studies, vertical flow systems, wherein influent water is pumped from the base of a column, tend to be preferred, as anaerobic conditions are easy to maintain and constant flow rates can be maintained.

3.2.1 Vertical flow systems

Different types of filter media have been examined in PRBs. Gilbert et al. (2008) studied seven types of materials (softwood, hardwood, coniferous, mulch, willow, compost and leaves) to select a suitable natural organic substrate to use in a PBR. Subsequent to a batch test, the material used in the laboratory-scale study was softwood. The columns were 0.09 m in diameter and 0.9 m long, and received an influent concentration of 50 mg NO_3-N dm^{-3} loaded from the column base at two HLRs: 0.3 cm^3 min^{-1} and 1.1 cm^3 min^{-1}. At the lower HLR, removals of more than 96% were measured, whereas removals of 66% were measured for the higher HLR. The impact of residence times was also studied by Claus and Kutzner (1985), who studied N removal in an up-flow packed bed reactor, with lava stones as support for the microbial growth. Using nitrate solution of different concentrations (1.8; 3.0; 4.3; 6.1 g NO^3 L^{-1}) and 5 different residence times (5; 3.3; 2.5; 2.0; 1.7 h), 95% denitrification was measured at the longest residence time.

Other types of filter media, such as shredded newspaper, have also been examined. Volokita et al. (1996) treated water in 0.55 m-high x 0.1 m-diameter laboratory columns using shredded newspaper (0.4 cm width). Complete nitrate removal of the inlet solution (100 mg L^{-1}) was achieved at an ambient temperature of 32°C. Sawdust has high denitrification rates due to its large surface area, but it is prone to clogging. Bedessem et al. (2005) used a mixture of sawdust and native soil in a 4.6 m-long, 7.6 cm-diameter laboratory column to treat synthetic wastewater. The total nitrogen (TN) removal was 31% in the control column (comprising only native soil) and 67% in columns with an organic layer (soil and sawdust). Saliling et al. (2007) evaluated woodchips and wheat straw using an up-flow bioreactor. The influent concentration was 200 mg NO_3-N L^{-1} and a 99% removal was obtained. Vrtovšek and Roš (2006) examined the effectiveness of a 1 m long x 0.12 m diameter fixed-bed biofilm reactor, comprising a mixture of PVC plastic and powdered activated carbon (PAC) as packing material. The reactor was inoculated with municipal wastewater before operation. Influent water with a concentration of 45 mg NO_3-N L^{-1} and sodium acetate ($CH_3COONa.3H_2O$) was loaded from the base of the column. Different loading rates were applied to the column, with drinking water quality being achieved at nitrogen loading rates (NLR) of lower than 1.9 g NO_3-N m^{-2} d^{-1}. Phillips and Love (2002) investigated a denitrifying bio-filter to remove nitrate from re-circulating aquaculture system waters using an up-flow fixed film column and two fermentation columns. Two nitrate concentrations (1.13 kg NO_3-N m^{-3} d^{-1} and 2.52 kg NO_3-N m^{-3} d^{-1}) were loaded at a HLR of 3.0 m hr^{-1}. The column was packed with polystyrene media with a specific surface area of 1000 m^2 m^{-3} and was seeded with activated sludge prior to operation. Commercial fish food was used as a fermentation source. Nitrate removal of greater than 99% was achieved. Rocca et al. (2007) used a coupling heterotrophic-autotrophic denitrification processes (HAD) supported by cotton and zero-valent iron (ZVI) to measure nitrate reduction. Two sets of columns filled with cotton and 150 g or 300 g of ZVI were used in this experiment. This had an up-flow inlet concentration of 100 and 220 mg NO_3 L^{-1}, and 3 and 6 mg L^{-1} of phosphate. The HAD had a higher volumetric nitrate removal ratio (VNR) than cotton-supported denitrification alone.

A laboratory sulphur-based reactive barrier system was evaluated by Moon et al. (2008) and was able to transform 60 mg N L^{-1} in di-nitrogen (N_2) in the presence of phosphate. The denitrification rate was higher than 95%. Cameron and Schipper (2010) compared nitrate removal, hydraulic and nutrient leaching characteristics of nine different carbon substrates. Mean nitrate removal rates for the period 10–23 months were 19.8 and 15 g N m^{-3} d^{-1} (maize cobs), 7.8 and 10.5 g N $m^{-3}d^{-1}$ (green waste), 5.8 and 7.8 g N m^{-3} d^{-1} (wheat straw), 3.0 and 4.9 g N m^{-3} d^{-1} (softwood), and 3.3 and 4.4 g N m^{-3} d^{-1} (hardwood) for the 14 and 23.5 C treatments, respectively.

3.2.2 Horizontal flow systems

Horizontal flow systems have also been used in studies. Healy et al. (2006) examined the use of various wood materials as a carbon source in laboratory horizontal flow filters to denitrify nitrate from a synthetic wastewater. The filter materials were: sawdust (*Pinus radiata*), sawdust and soil, sawdust and sand, and medium-chip woodchips and sand. Two influent NO_3-N concentrations, 200 mg L^{-1} and 60 mg L^{-1}, loaded at 2.9 to 19.4 mg NO_3-N kg^{-1} mixture d^{-1}, were used. The horizontal flow filter with a woodchip/sand mixture, loaded at 2.9 mg NO_3-N kg^{-1} d^{-1}, performed best, yielding a 97% reduction in NO_3-N at steady-state conditions. Using a sand tank containing a denitrifying zone in the centre (sand coated with soybean oil), Hunter (2001) measured a 39% nitrate removal of the initial concentration of 20 mg NO_3-N L^{-1} at a flow rate of 1112 L $week^{-1}$.

3.3 Wetlands

Dairy dirty water (DDW) can have a significant adverse effect on the environment. In Ireland, but in recent years, the use of constructed wetlands (CWs) for the treatment of DDW, as well as domestic and municipal wastewaters, has being gaining in popularity. This is due to their relatively low capital costs and maintenance requirements.

3.3.1 Wetland types

There are two types of CW: free water surface constructed wetlands (FWS CWs) and subsurface CWs. In FWS CWs, wastewater flows in a shallow water layer over a soil substrate. Subsurface CWs may be either subsurface horizontal flow CWs (SSHF CWs) or subsurface vertical flow CWs (SSVF CWs). In SSHF CWs, wastewater flows horizontally through the substrate. In SSVF CWs, wastewater is dosed intermittently onto the surface of sand and gravel filters and gradually drains through the filter media before collecting in a drain at the base. CWs may be planted with a mixture of submerged, emergent and, in the case of FWS CWs, floating vegetation. However, the ability of vegetation to capture nutrients, particularly in a cool temperate climate, is limited (Healy et al., 2007).

The large surface area of CWs provides an environment for the physical/physico-chemical retention and biological reduction of organic matter and nutrients (Knight et al., 2000; Lu et al., 2009). Depending on the type of CW used, its design, organic loading rate (OLR) and hydraulic retention time (HRT) (Karpiscak et al., 1999); a CW can have a significant nutrient removal capability. However, due to the effect of changing temperatures, the treatment efficiency of these systems tends to vary throughout the year (Bachand and Horne, 2000).

3.3.2 Design guidelines for dairy dirty water treatment

American guidelines for the design loading of SSHF CWs treating agricultural wastewater (NRCS, 1991) recommend an areal OLR of 7.3 g 5-day biochemical oxygen demand (BOD_5) $m^{-2}d^{-1}$; similar rates are used in wetland design for cool temperate climates (Cooper et al., 1996; Dunne et al., 2005ab). New Zealand guidelines for the disposal of DDW (Tanner and Kloosterman, 1997) recommended that an FWS CW should only succeed two waste stabilization ponds (an anaerobic and an aerobic pond, respectively) before entering the wetland with an OLR not exceeding 3 g BOD_5 m^{-2} d^{-1}. Generally, FWS CWs are used for the treatment of DDW as issues such as blockage of the filter media – normally associated with the operation of SSHF CWs – do not arise.

3.3.3 Treatment efficacy

Results from CWs have been variable. Table 4 tabulates the performance of FWS CWs in the treatment of DDW in a range of countries. In a study of planted and unplanted SSHF CWs, where the unplanted SSHF CWs acted as an experimental control, Tanner (1995ab) found that under 5-day carbonaceous biochemical oxygen demand ($CBOD_5$) OLRs ranging from 0.9 to 3.4 g $CBOD_5$ m^{-2} d^{-1} (unplanted) and 0.9 to 4.1 g $CBOD_5$ m^{-2} d^{-1} (planted), maximum $CBOD_5$ removals of 85% and 92%, respectively, were measured. Ammonification was more pronounced with increasing HRT, and total nitrogen (Tot-N) removal varied between 48 and 80% for planted CWs. Similar OLRs were used in a study on a 3-cell integrated FWS CW in Co. Wexford, Ireland (Dunne et al., 2005ab) where, under OLRs varying from 2.7 to 3.5 g BOD_5 m^{-2} d^{-1}, good organic removal was measured, but nitrification was not complete during winter.

Cronk et al. (1994) also found that under reduced retention times (with OLRs of 60 g BOD_5 m^{-2} d^{-1}) BOD_5 and suspended solids (SS) concentrations were not reduced to acceptable levels after treatment in a 1-cell FWS CW, and that no significant reduction of total kjeldahl nitrogen (TKN) occurred. In a study on a dairy farm in Drointon in the U.K (Cooper et al., 1996), a SSHF CW was used to treat influent with an average BOD_5 concentration of 1192 mg L^{-1}. The system initially utilized only the wetland alone and performed poorly under an OLR of approximately 26 g BOD_5 m^{-2} d^{-1}. However, when two SSVF CWs and a lagoon were

Parameter	Wetland Type	Loading rate	Influent ± SD	Effluent ± SD	Removal efficiency	Reference
BOD						
Ireland	FWS		998±1034	16±5	98	[1]
USA	FWS	~60	7130	2730	62	[2]
	FWS	~12	242	246	-2	[3]
	FWS	NP	1914	59	97	[4]
	FWS	18	2680	611	77	[5]
Australia	FWS	5.6	220	90	59	[6]
Italy	FWS	~1.9	451	28	94	[7]
N. Zealand	FWS	~4.1	113	27	76	[8]
	FWS	~1	337	11	92	[8]
COD						
Ireland	FWS		1718±2008	162±83	91	[1]
SS						
Ireland	FWS		535±434	34±31	94	[1]
USA	FWS	NP	5540	990	82	[2]
	FWS	NP	911	641	30	[3]
	FWS	NP	1645	65	96	[4]
	FWS	9	1284	130	90	[5]
N. Zealand	FWS	~8.5	150	33	78	[8]
	FWS	~1.9	142	34	76	[8]
Tot-N						
USA	FWS	0.7	103	74	28	[5]
	FWS	NP	170	13	92	[4]
N. Zealand	FWS	2.7	~38	20	48	[9]
	FWS	0.6	~38	10	75	[9]
NH$_4$-N						
Ireland	FWS		48±25	6±5	88	[1]
USA	FWS	0.05	8	52	0	[5]
	FWS	NP	72	32	56	[4]
Israel	FWS	NP	51	44	14	[10]
N. Zealand	FWS	NP	33	22	34	[9]
	FWS	NP	38	11	71	[9]
NO$_3$-N						
USA	FWS	NP	5.5	10	0	[4]
	FWS	2x10^{-3}	0.3	0.1	67	[5]
Tot-P						
USA	FWS	NP	53	2.2	96	[4]
	FWS	0.2	26	14	46	[5]
N. Zealand	FWS	0.8	~11	6.9	37	[9]
	FWS	0.2	~11	2.9	74	[9]
PO$_4$-P						
Ireland	FWS		15±7	3±2	80	[1]

Avg±SD; FWS = free-water surface constructed wetland; NP = not published
[1] Healy and O' Flynn (pers. comm.); [2] Cronk et al., 1994; [3] Karpiscak et al., 1999; [4] Schaafsma et al., 2000; [5] Newman et al., 2000; [6] Geary and Moore, 1999; [7] Mantovi et al., 2003; [8] Tanner et al., 1995(a); [9] Tanner et al., 1995 (b); [10] Ran et al., 2004.

Table 4. Average influent and effluent concentration (mg L^{-1}), loading rates (g m^{-2} d^{-1}), and removal efficiencies of wetlands treating dairy dirty water (DDW).

installed in front of the SSHF CW, the system had an OLR of approximately 4 g BOD_5 m^{-2} d^{-1} and had good organic and SS removal rates, but had limited nitrification due to large fluctuations in the inlet wastewater strength. Even under significantly reduced OLRs, SSHF and FWS CWs have under-performed. In Italy, a study on a 2 cell FWS CW operated in series and monitored over a 26 month period, treating a mixture of domestic and DDW at an average influent OLR under 2 g BOD_5 m^{-2} d^{-1}, showed that anoxic zones which developed in the wetland inlet meant that nitrification was inhibited, producing an effluent Tot-N which was mainly composed of ammonium-N (NH_4-N) (Mantovi et al., 2003).

Present agricultural practice in Ireland is governed by The European Communities (Good Agricultural Practice for Protection of Waters) Regulations 2009 (S.I. No. 101 of 2009), which places a responsibility on the individual farmer and the public authority to adhere to the conditions set out within the Nitrates Directive (EEC, 1991(a)) and other water quality directives to ensure good wastewater management practices. On account of this, CWs are becoming popular for the treatment of DDW. Healy and O' Flynn (*pers. comm.*) evaluated the performance of seven CWs treating DDW in Ireland. They found that average removals of chemical oxygen demand (COD) from DDW were 91%. However, average effluent concentrations were 162 mg L^{-1}, which was much higher than the maximum allowable concentration (MAC). The performance of the CWs in the reduction of NH_4-N and ortho-phosphorus (PO_4-P) was also highly variable.

4. Conclusion

Much research focuses on the nutrient content of agricultural wastewaters and their inorganic fertilizer replacement potential. Many options for dairy slurry and dirty water are in place including land application, irrigation and treatment using a variety of on farm or off farm options. Nutrient, gaseous and microbial losses can result from land application of agricultural wastes. Much research focuses on matching crop requirements with organic fertilizer applications. In addition, the control of P within such wastes can prevent incidental losses to the environment e.g. chemical amendments. Once nutrients are lost, other forms of remediation such as PRB's or wetlands may be applicable to protect a waterbody.

5. References

ADAS, 1985. Dirty water disposal on the farm, Ministry of Agriculture, Fisheries and Food, London.

ADAS, IGER & SSRC, 1994. Low rate irrigation of dilute farm wastes. Report for the National Rivers Authority, R and D no. 262, National Rivers Authority, Bristol.

Aislabie, J.M., Fraser, R.H., Smith, J. & McLeod, M. 2001. Leaching of bacterial indicators through four New Zealand soils. Australian Journal of Soil Research 39: 1397-1406.

Anon, 2005. European Communities (Good Agricultural Practice for Protection of Waters) Regulations 2005. SI 788 of 2005. Department of Environment, Heritage and Local Government, The Stationary Office, Dublin, 47 pp.

Anon, 2006. European Communities (Good Agricultural Practice for Protection of Waters) Regulations 2006. SI 378 of 2006. Department of Environment, Heritage and Local Government, The Stationary Office, Dublin, 49 pp.

Anon, 2009. European Communities (Good Agricultural Practice for Protection of Waters) Regulations 2009. SI 101 of 2009. Department of Environment, Heritage and Local Government, The Stationary Office, Dublin, 50 pp.

Anon, 2010. European Communities (Good Agricultural Practice for Protection of Waters) Regulations 2010. SI 610 of 2010. Department of Environment, Heritage and Local Government, The Stationary Office, Dublin, 53 pp.

Bachand, P.A.M. & Horne, A.J. 2000. Denitrification in constructed wetland free-water surface wetlands:II. Effects of vegetation and temperature. Ecological Engineering 14: 17-32.

Baudart, J. Grabulos, J., Barusseau, J.P. & Lebaron, P. 2000. Salmonella spp. and fecal coliform loads in coastal waters from a point vs. non-point source of pollution, Journal of Environmental Quality 29: 241–250

Bedessem, M.E., Edgar, T.V. & Roll, R. 2005. Nitrogen Removal in Laboratory Model Leachfields with Organic-Rich Layers. Journal of Environmental Quality 34(3): 936-942.

Bolan, N. & Swain, D. 2004. Issues and innovations in land application of farm wastes. New Zealand Journal of Agricultural Research 47: 387-388.

Brennan, F. P., O'Flaherty, V., Kramers, G., Grant, J. & Richards, K.G. 2010. Long-term Persistence and leaching of Escherichia coli in Temperate Maritime Soils. Applied and Environmental Microbiology 76 (5): 1449–1455.

Brennan, R.B., Fenton, O. & Healy, M.G. 2010a. Chemical amendment of dairy cattle slurry for control of phosphorus in runoff from grasslands. A05 symposium on emerging technologies to remove phosphorus. ASA-CSA-SSSA International Meetings, Long Beach, CA. 31st October – 4th November, 2010.

Brennan, R.B., Fenton, O. & Healy, M.G. 2010b. Evaluation of chemical amendments to control soluble phosphorus losses from dairy cattle slurry. In: E. Turtola, P. Ekholm, and W. Chardon (Eds). MTT Science 10. Novel methods for reducing agricultural nutrient loading and eutrophication: Meeting of COST 869, 14th -16th June, Jokioinen, Finland. p. 18.

Brennan, R.B., Fenton, O., Rodgers & M., Healy, M.G. 2010c. The addition of chemical amendments to dairy cattle slurry for the control of phosphorus in runoff from grasslands. BSAS/WPSA/Agricultural Research Forum Conference. Queen's University, Belfast. 12th -14th April, 2010.

Cameron, S.G. & Schipper, L.A. 2010. Nitrate removal and hydraulic performance of organic carbon for use in denitrification beds. Ecological Engineering 36(11): 1588-1595.

Carlyle, J.C., Bligh, M.W. & Nambiar, E.S. 1998. Woody residue management to reduce nitrogen and phosphorus leaching from sandy soil after clear-felling *Pinus radiata* plantations. Canadian Journal of Forest Research 28(8): 1222-1232.

Claus, G. & Kutzner, H.J. 1985. Autotrophic denitrification by Thiobacillus denitrificans in a packed bed reactor. Applied Microbiology and Biotechnology 22(4): 289-296.

Collins, R., McLoed, M., Donnison, A., Close, M., Hanly, J., Horne, D., Ross, C., Davies-Colley, R., Bagshaw, C. & Mattews, L. 2007. Best management practices to mitigate faecal contamination by livestock of New Zealand waters. New Zealand Journal of Agricultural Research 50: 267-278.

Cooper, P., Job, G.D., Green, M.B. & Shutes, R.B.E. 1996. Reed beds and constructed wetlands for wastewater treatment. WRc, Swindon, U.K, 184 pp.

Craggs, R.J., Sukias, J.P. & Davies-Colley, R.J. 2004. Advanced pond system for dairy-farm effluent treatment. New Zealand Journal of Agricultural Research 47: 449-460.

Cronk, J.K, Kodmur, V. & Shirmohammadi, A. 1994. An evaluation of wetlands for the treatment of dairy effluent: results from the first year of operations. Presented at the International winter meeting of the American Society of Agricultural Engineers. Pap. No. 94-2600, ASAE, St. Joseph, MI, USA, pp. 11-19.

CSO, 2006. Livestock survey, February 2006.

Culleton, N., Liebhardt, W.C., Murphy, W.E., Cullen, J. & Cuddity, A. 2000. Thirty years of phosphorus fertilizer on Irish pastures: animal-soil-water relationships, Teagasc, Wexford.

DAFF, 1996. Code of good agricultural practice to protect waters from pollution by nitrates.

DAFF, 2003. Compendium of Irish agricultural statistics, Department of Agriculture and Food., Wexford.

Dao, T. H. 1999. Co-amendments to modify phosphorus extractibility and nitrogen/phosphorus ration in feedlot manure and composted manure. Journal of Environmental Quality 28:1114-1121.

Demanet, R., Aguilera, M. & Mora, M. 1999. Efecto de la aplicacion de prunes sobre el sistema suela-planta. Frontera Arricola No 5, 87-94.

Donnison, A. & Ross, C. 2003. Campylobacter and farm dairy effluent irrigation. New Zealand Journal of Experimental Agriculture 46:255-262

Dou, Z., Zhang, G. Y., Stout, W. L., Toth, J. D. & Ferguson, J. D. 2003. Efficacy of Alum and Coal Combustion By-Products in Stabilizing Manure Phosphorus. Journal of Environmental Quality 32: 1490-1497.

Dunne, E.J., Culleton, N., O'Donovan, G., Harrington, R. & Olsen, A.E. 2005a. An integrated constructed wetland to treat contaminants and nutrients from dairy farmyard dirty water. Ecological Engineering 24: 221-234.

Dunne, E.J., Culleton, N., O'Donovan, G., Harrington, R. & Daly, K., 2005b. Phosphorus retention and sorption by constructed wetland soils in Southeast Ireland. Water Research 39(18): 4355-4362.

EC, 1991. Directive 91/676/EEC concerning the protection of waters against pollution caused by nitrates from agricultural sources. 375: 1-8.

EC, 1998. Council Directive (98/83/EC) of 3rd November 1988 on the quality of water intended for human consumption. Official journal of the European Communities L220/32. Brussels, Belgium.

EC, 2000. Water Framework Directive (2000/60/EC) establishing a framework for community action in the field of water policy.

EEC, 1975. Council Directive (75/440/EEC) of 16th of June 1975 concerning the quality required of surface water intended for the abstraction of drinking water in the Member States. . Official Journal of the European Communities.

EEC, 1980. Council Directive (80/68/EEC) of 17th December 1979 on the protection of groundwater against pollution caused by certain dangerous substances. Official Journal of the European communities L20: 43-48.

EEC, 1991a. Council Directive (91/271/EEC) of 12th December 1991 concerning the protection of waters caused by nitrates from agriculture sources. Official journal of the European Communities.

EEC, 1991b. Council Directive (91/271/EEC) of 21st May 1991 concerning urban wastewater treatment. Official Journal of the European Communities.

EPA, 2004. Biological survey of river quality - results of the 2003 investigations EPA, Wexford.

EPA, 2008. Water Quality in Ireland 2004 - 2006. Available at: http://www.epa.ie/downloads/pubs/water/waterqua/waterrep/#d.en.25320 [Accessed November 8, 2010].

Fenton, O., Healy, M. & Schulte, R.P.O., 2008. A review of remediation and control systems for the treatment of agricultural waste water in Ireland to satisfy the requirements of the Water Framework Directive. Biology and Environment Proceedings of the Royal Irish Academy 108(B): 69-79.

Fenton, O., Healy, M.G. & Rodgers, M., 2009a. Use of ochre from an abandoned metal mine in the South East of Ireland for phosphorus sequestration from dairy dirty water. Journal of Environmental Quality 38(3): 1120-5.

Fenton, O., Richards, K.R., Haria, A., Johnson, P., Horan, B., Murphy, P., Premrov, A., Coxon, C., Humphreys, J. & Necpalova, M. 2009.Travel time from source to receptor in four contrasting case studies in Ireland to illustrate the effect of lag time on achieving the goals of the WFD. Tearmann: The Irish Journal of Agri-Environmental Research 7: 129-142.

Fingelton, W.A. and Cushion, M., 1999. Irish agriculture in Figures 1998, Teagasc, Dublin.

Food Safety Authority of Ireland. 2008. Food safety implications of land-spreading agricultural, municipal and industrial organic materials on agricultural land used for food production in Ireland, FSAI

Geary, P.M. & Moore, J.A. 1999. Suitability of a treatment wetland for dairy wastewaters. Water Science and Technology 40 (3): 179-185.

Ghani A, Dester M, Sarathchandra U. & Waller J. 2005. Effects of dairy factory effluent application on nutrient transformation in soil. New Zealand Journal of Agricultural Research 48: 241–253.

Gibert, O. 2008. Selection of organic substrates as potential reactive materials for use in a denitrification permeable reactive barrier (PRB). Bioresource Technology 99(16): 7587-7596.

Hawke, R.M. & Summers, S.A. 2006. Effects of land application of farm dairy effluent on soil properties: a literature review New Zealand Journal of Agricultural Research 49: 307-320.

Healy, M.G., Newell, J. & Rodgers, M. 2007. Harvesting effects on biomass and nutrient retention in a free-water surface constructed wetland in western Ireland. Biology and Environment: Proceedings of the Royal Irish Academy 107B: 139-145.

Healy, M.G., Rodgers, M. & Mulqueen, J. 2006. Denitrification of a nitrate-rich synthetic wastewater using various wood-based media materials. Journal of Environmental Science and Health. Part A, Toxic/Hazardous Substances and Environmental Engineering 41(5): 779-788.

Healy, M., Rodgers, M. & Mulqueen, J. 2007. Treatment of dairy wastewater using constructed wetlands and intermittent sand filters. Bioresource Technology 98(12): 2268-2281.

Hiscock, K.; Lovett, A.; Saich, A.; Dockerty, T.; Johnson, P.; Sandhu, C.; Sunnenberg, G.; Appleton, K.; Harris, B. & Greaves, J. 2007. Modelling land-use scenarios to reduce groundwater nitrate pollution: the European Water4All project. Quarterly Journal of Engineering Geology and Hydrogeology 40(4): 417-434.

Houlbrooke, D.J., Horne, D.J., Hedley, M.J., Hanly, J.A., Scotter, D.R. & Snow, V.O. 2004. Minimising surface water pollution from farm dairy effluent application to mole pipe drained soils. An evaluation of the deferred irrigation system for sustainable land treatment in the Manawatu. New Zealand Journal of Agricultural Research 47: 405-415.

Houlbrooke, D.J., Horne, D.J., Hedley, M.J., Hanly, J.A., D.R. & Snow, V.O. 2004. A review of literature on the land treatment of farm dairy effluent in New Zealand and its impacts on water quality. New Zealand Journal of Agricultural Research 47: 499-511.

Humphreys, J., Casey, I.A., Darmody, P., O Connell, K., Fenton, O. & Watson, C.J. 2008. Quantities of mineral N in soil and concentrations of nitrate-N in groundwater in four grassland-based systems of dairy production on a clay-loam soil in a moist temperate climate. Grass and Forage Science 63(4): 481-494.

Hunter, W.J. 2001. Use of vegetable oil in a pilot-scale denitrifying barrier. Journal of Contaminant Hydrology 53(1-2): 119-131.

Jamieson, R. C., Gordon, R.J., Sharples, K.E., Stratton, G. W. & Madani, A. 2002. Movement and persistence of faecal bacteria in agricultural soils and subsurface drainage water: a review. Canadian Biosystems Engineering 44:1.1–1.9.

Kalbasi, M. & Karthikeyan, K. G. 2004. Phosphorus dynamics in soils receiving chemically treated dairy manure. Journal of Environmental Quality 33: 2296-2305.

Karpiscak, M.M., Freitas, R.J., Gerba, C.P., Sanchez, L.R. & Shamir, E. 1999. Management of dairy waste in the Sonoran Desert using constructed wetland technology. Water Science and Technology 40 (3): 57-65.

Kay, D., Aitken, M., Crowther, M.J., Dickson, I., Edwards, A.C., Francis, C., Hopkins, M., Jeffrey, W., Kay, C., McDonald, A.T., McDonald, D., Stapleton, C.M., Watkins, J., Wilkinson, J. & Wyer, M.D. 2007. Reducing fluxes of faecal indicator compliance parameters to bathing waters from diffuse agricultural sources: The Brighouse Bay study, Scotland. Environmental Pollution 147: 138-149.

Kebede-Westhead, E., Pizarro, C. & Mulbry, W.W. 2004. Treatment of dairy manure effluent using freshwater algae: elemental composition of algal biomass at different manure loading rates. Journal of Agricultural Food Chemistry 52(24): 7293-6.

Knight, R.L., Payne Jr., V.W.E., Borer, R.E.,Clarke Jr., R.A. & Pries, J.H. 2000. Constructed wetlands for livestock management. Ecological Engineering 15: 41-55.

Lalor, S.T.J., Schröder, J.J., Lantinga, E.A., Oenema, O., Kirwan, L & Schulte, R.P.O. 2010 (a). Nitrogen fertilizer replacement value of cattle slurry in grassland as affected by method and timing of application. Journal of Environmental Quality. Published online September 2010.

Lalor, S. 2010(b). Efficient and reliable utilisation of nutrients in animal manures. Technical report, Teagasc, 2010.

Lu, S.Y., Wu, F.C., Lu, Y.F., Xiang, C.S., Zhang, P.Y. & Jin, C.X. 2009. Phosphorus removal from agricultural runoff by constructed wetland. Ecological Engineering 35: 402-409.

Lucey, J. 2009. Water quality in Ireland 2007 – 2008. Key indicators of the aquatic environment. EPA, Wexford.

MAFF, 2000. Fertilizer recommendations handbook for agriculture and horticulture crops (RB209). 7th ed. 175 p. Ministry of Agriculture, Fisheries and Food (MAFF). The Stationary Office, London, U.K.

Mantovi, P., Marmiroli, M., Maestri, E., Tagliavini, S., Piccinini, S. & Marmiroli, N. 2003. Application of a horizontal subsurface flow constructed wetland on treatment of dairy parlour wastewater. Bioresource Technology 88: 85-94.

Martínez -Suller, L. Provolo, G., Brennan, D., Howlin, T., Carton, O.T., Lalor, S.T.J. & Richards, K.G. 2010a. A note on the estimation of nutrient value of cattle slurry using easily determined physical and chemical parameters. Irish Journal of Agricultural and Food Research 49: 93-97.

Martínez -Suller, L. Provolo, G., Carton, O.T., Brennan, D., Kirwan, L. & Richards., K.G. 2010b. The composition of dirty water on dairy farms in Ireland. Irish Journal of Agricultural and Food Research 49: 67-80.

McFarland, A. M. S., Hauck, L. M. & Kruzic, A. P. 2003. Phosphorus reductions in runoff and soils from land-applied dairy effluent using chemical amendments: An observation. The Texas Journal of Agriculture and Natural resource, 47-59.

McLeod, M., Aislabie, J.M., Ryburn, J., McGill, A. & Taylor, M.D. 2003. Microbial and chemical tracer movement through two Southland soils, New Zealand. Australian Journal of Soil Research 41: 1163-1169.

Minogue, D., Murphy, P., French, P., Coughlan, F. & Bolger, T., 2010. Characterisation of soiled water on Irish dairy farms. , BSAS & Agricultural Research Forum, Belfast, N.Ireland.

Moon, H.S. 2008. A long-term performance test on an autotrophic denitrification column for application as a permeable reactive barrier. Chemosphere 73(5): 723-728.

Moore, P. A., Daniel, T. C., Gilmour, J. T., Shreve, B. R., Edwards, D. R. & Wood, B. H. 1998. Decreasing metal runoff from poultry litter with aluminium sulphate. Journal of Environmental Quality 27: 92-99.

Moore, P. A. & Edwards, D. R. 2007. Long-term effects of poultry litter, alum-treated litter, and ammonium nitrate on phosphorus availability in soils. J Environ Qual, 36, 163-174.

Mulbry, W.W. & Wilkie, A.C., 2001. Growth of benthic freshwater algae on dairy manures. Journal of Applied Phycology 13: 301-306.

Newman, J.M., Clausen, J.C. & Neafsey, J.A. 2000. Seasonal performance of a wetland constructed to process dairy milkhouse wastewater in Connecticut. Ecological Engineering 14: 181-198.

Novak, J. M. & Watts, D. W. 2005. An alum-based water treatment residual can reduce extractable phosphorus concentrations in three phosphorus-enriched coastal plain soils. Journal of Environmental Quality 34: 1820-1827.

NRCS. 1991. Technical requirements of agricultural wastewater treatment. National Bulletin No. 210-1-17. National Resource Conservation Service, United States Department of Agriculture, Washington, DC, USA.

Pell, A.N. 1997. Manure and microbes: public and animal health problem? Journal of Dairy Science 80: 2673-2681.

Penn, C. J., Bryant, R. B., Callahan, M. A. & McGrath, J. M. 2009. Use of industrial byproducts to sorb and retain phosphorus. Communications in Soil Science and Plant Analysis, In print.

Phillips, J.B. & Love, N.G. 2002. Biological denitrification using upflow biofiltration in recirculating aquaculture systems: Pilot-scale experience and implications for full-scale. In Proceedings of the 2nd International Conference on Recirculating Aquaculture.

Pizarro, C., Kebede-Westhead, E. & Mulbry, W. 2002. Nitrogen and phosphorus removal rates using small algal turfs grown with dairy manure. Journal of Applied Phycology 14(6): 469-473.

Provolo, G. & Martínez -Suller, L. 2007. In situ determination of slurry nutrient content by electrical conductivity. Bioresource Technology 98: 3235-3242.

Ran, N., Agami, M. & Oron, G. 2004. A pilot study of constructed wetlands using duckweed (Lemma gibba L.) for treatment of domestic primary effluent in Israel. Water Research 38: 2241 – 2248.

Regan, J.T., Rodgers, M., Healy, M.G., Kirwan, L. & Fenton, O. 2010. Determining phosphorus and sediment release rates from five Irish tillage soils. Journal of Environmental Quality 39(1): 185-92.

Richards, K., Ryan, M. & Coxon, C.E. 2004. Dirty water - A valuable source of nitrogen on dairy farms, 12th Nitrogen Workshop, Reading, U.K. In Controlling Nitrogen flows and losses, Hatch, D.J., Chadwick, D.R., Jarvis, S.C. and Roker, J.A. (Eds) Wageningen Academic Publishers, The Netherlands: 439-440.

Rocca, C.D., Belgiorno, V. & Meriç, S. 2007. Heterotrophic/autotrophic denitrification (HAD) of drinking water: prospective use for permeable reactive barrier. Desalination, 210(1-3): 194-204.

Ryan, D., 2005. A slurry spreader to meet farming needs and environmental concerns. Teagasc, Wexford.

Salazar, F., Carlos-Dumont, J., Chadwick, D., Salana, R. & Santana, M. 2007. Characterisation of dairy slurry in southern Chile farms. Agricultura Technica (Chile) 67(2): 155-162

Saliling, W.J.B., Westerman, P.W. & Losordo, T.M. 2007. Wood chips and wheat straw as alternative biofilter media for denitrification reactors treating aquaculture and other wastewaters with high nitrate concentrations. Aquacultural Engineering, 37(3): 222-233.

Sapkota, A.R., Curriero, F.C., Gibson, K.E. & Schwab, K.J. 2007. Antibiotic-resistant enterococci and fecal indicators in surface water and groundwater impacted by a concentrated swine feeding operation. Environmental Health Perspectives 115: 1040-1045.

Schulte, R., Melland, A., Richards, K., Fenton, O. & Jordan, P. 2010. Modelling soil phosphorus decline; expectations of Water Framework Directive policies in Ireland. Environmental Science and Policy 13: 472-484.

Scotford, I.M., Cumby, T.R., Han, L. & Richards, P.A. 1998a. Development of a prototype nutrient sensing system for livestock slurries. Journal of Agricultural Engineering Research 69: 217-228.

Scotford, I.M., Cumby, T.R., White, R.P. Carton, O., Lorenz, F., Hatterman, U. & Provolo, G. 1998b. Estimation of the nutrient value of agricultural slurries by measurement of physical and chemical properties. Journal of Agricultural Engineering Research 71: 291-305.

Schaafsma, J.A., Baldwin, A.H. & Streb, C.A. 2000. An evaluation of a constructed wetland to treat wastewater from a dairy farm in Maryland, USA. Ecological Engineering 14: 199-206.

Schröder, J.J., Jansen, A.G. & Hilhorst, G.J. 2005. Long-term nitrogen fertilizer value of cattle slurry. Soil Use and Management 21: 196-204.

Serrenho, A.J., Fenton, O., Rodgers, M. & Healy, M.G. 2010. Laboratory study of a denitrification system using a permeable reactive barrier. BSAS/WPSA/Agricultural Research Forum Conference. Queen's University, Belfast. 12-14 April, 2010.

Serrenho, A., Fenton, O., Lanigan, G. & Healy, M.G. 2010. Pollution swapping in four laboratory carbon rich denitrification systems. In: Raghuram N., Bijay-Singh, Pathak H., Ahmad A., Jain V. and Pal M. (Eds) Proceedings of the 5th International Nitrogen Conference "Reactive nitrogen management for sustainable development – science, technology and policy" 3-7th December 2010 Ashok Hotel, New Dehli, India, p.272.

Skerrett H.E. & Holland, C.V. 2000. The occurrence of Cryptosporidium in environmental waters in the greater Dublin area, Water Research 34: 3755–3760

Smith, D. R., Moore, P. A., Jr., Haggard, B. E., Maxwell, C. V., Daniel, T. C., Van Devander, K., Davis, M. E. 2004. Effect of aluminum chloride and dietary phytase on relative ammonia losses from swine manure. Journal of Animal Science 82: 605-611.

Smith, K. & Chambers, B. 1993. Utilising the nitrogen content of organic manures on farms-problems and practical solutions. Soil Use and Management 9: 105-112.

Sommer S.G., Schjoerring J.K., Denmead O.T. 2004. Ammonia emission from mineral fertilizers and fertilized crops. Advances in Agronomy 82, 557-622

Stark, C.H. & Richards, K.G. 2008. The Continuing Challenge of Agricultural Nitrogen Loss to the Environment in the Context of Global Change and Advancing Research. Dynamic Soil, Dynamic Plant 2: 1-12.

Tanner, C.C. & Kloosterman, V.C. 1997. Guidelines for constructed wetland treatment of farm dairy wastewaters in New Zealand. NIWA Science and Technology Series No. 48. Hamilton, New Zealand.

Tanner, C.C., Clayton, J.S. & Upsdell, M.P. 1995a. Effect of loading rate and planting on treatment of dairy farm wastewaters in Constructed Wetlands-I. removal of oxygen demand, suspended solids and faecal coliforms. Water Research 29: 17-26.

Tanner, C.C., Clayton, J.S. & Upsdell, M.P. 1995b. Effect of loading rate and planting on treatment of dairy farm wastewaters in constructed wetlands-II. removal of nitrogen and phosphorus. Water Research 29: 27-34.

Topp, E., Scott, A., Lapen, D.R., Lyautey, E. & Duriez, P. 2009. Livestock waste treatment systems for reducing environmental exposure to hazardous enteric pathogens: Some considerations. Bioresource Technology 100: 5395-5398.

Tunney, H., Coulter, B., Daly, K., Kurz, I., Coxon, C., Jeffery, D., Mills, P., Kiely, G. & Morgan, G. 2000. Quantification of phosphorus loss to water due to soil P desorption. Final report. Environmental Monitoring R&D Report Series No.6. EPA, Wexford.

Tyrrel, S.F. 1999. The microbiological quality of water used for irrigation, Irrigation News 27, pp. 39–42.

UNEP, 2003. Freshwater in Europe. Available at: http://www.grid.unep.ch/product/publication/freshwater_europe/sum.php [Accessed November 7, 2009].

Vanotti, M.B., Szogi, A.A., Hunt, P.G., Millner, P.D. & Humenik, F.J., 2007. Development of environmentally superior treatment system to replace anaerobic sine lagoons in the USA. Bioresource Technology 98:3184–3194.

Venglovsky, J., Sasakova, N. & Placha, I. 2009. Pathogens and antibiotic residues in animal manures and hygienic and ecological risks related to subsequent land application. Bioresource technology 100:5386-5391

Villar, M.C., Diaz-Fierros, F., Cabaneiro, A., Leiros, M., Gil Sostres, F., Carballas, M. & Carballas, T. 1979. Metodos repidos para la caracterizacion del purin de vacuno. In: "Aprovechamiento de maeriales organicos residuals como fertilzantes en la agricultura gallega" Project N.3365-79.

Volokita, M., Belkin, S., Abeliovich, A. & Soares, M.I.M. 1996. Biological denitrification of drinking water using newspaper. Water Research 30: 965-971.

Vrtovšek, J. & Roš, M. 2006. Denitrification of Groundwater in the Biofilm Reactor with a Specific Biomass Support Material. Acta Chim. Slov, 53: 396–400.

Wang L, Li Y, Chen P, Min M, Chen Y, Zhu J. & Ruan RR. 2010. Anaerobic digested dairy manure as a nutrient supplement for cultivation of oil-rich green microalgae Chlorella sp. Bioresource Technology 101 (8): 2623–2628.

Ward, A.J., Hobbs, P.J., Holliman, P.J. & Jones, D.L. 2008. Optimisation of the anaerobic digestion of agricultural resources. Bioresource Technology 99: 7928-7940.

Water4all, 2005. Sustainable Groundwater Management, Handbook of best practice to reduce agricultural impacts on groundwater quality.

7

Water Quality of Streams Receiving Municipal Waste Water in Port Harcourt, Niger Delta, Nigeria

Alex C. Chindah[1], Solomon A. Braide[1], and Charles C. Obunwo[2]
[1]*Institute of Pollution Studies*
[2]*Chemistry Department*
Rivers State University of Science and Technology, PMB 5080, Port Harcourt,
Nigeria

1. Introduction

The Niger Delta environment was relatively a pristine area some 100 years ago and consists of several ecological zones mainly lowland forest, freshwater swamp forest, prominent in the northern limit while the mangrove and barrier island zones dominate the southern stretch (RPI, 1985, NDES, 2000 and NDDC, 2004). Settlements were of small population and largely in pockets around these ecological zones. The people are agrarian and indulge mostly in farming, fishing and exploitation of timber and non timber forest products. With the relative small nature of the populations in the settlements their wastes generated and discharged into the environment had little or no significant impact on the environment (Onuoha, *et. al.*, 1991).

With the absence of pipe borne water they depended on the stream system for the potable water use, recreation, washing, bathing and fishing (Amadi *et. al.*, 1997).

The advent of civilization has attracted human population to the major urban centres for white collar jobs and more also the crude oil found in commercial quantity in the region has accelerated the pace of development in terms of human population, urban growth, industrial activities, infrastructural development, intensive farming and other economic activities (NDDC,2004, Petrarova *et. al.*, 2009, Onderka *et. al.*, 2010).

The growth of human population and rapid industrialization led to increasing use of urban waters as sewers, compromising their other uses. The discharge of industrial effluents has led inevitably, to alterations in the quality and ecology of receiving water bodies (Sheikh, and Irshad. 1980 and Wahid *et. al.*, 1999). This results into new challenges to water resource managers and aquatic ecologists. Several attempts have been made to regulate/control the quality of effluents that are discharged from waste generating industries into the water systems with little effort on urban discharges. Today, most urban areas of the developing world remain inadequately served by sewage treatment infrastructure (NDDC,2004). Untreated wastes pose serious threats to associated environment including human health risks. Commonly cited effects of industrial effluents on the receiving waters are high turbidity, reduced transparency, increased suspended solids and oxygen depletion (Rafiu *et. al.*, 2007). The area study covers over 94.72 km^2 with a population of about 1.9 million.

The tremendous spatial spread of the Port Harcourt city has resulted in land take for various purposes, encroached, and converted to build up area with concrete buildings by both the government and private agencies without providing open spaces and corridors. Consequently, the natural water bodies (stream) draining the forest seawards are left bare including the stream banks especially in Port Harcourt area where this study was undertaken. As a result the water bodies now lack ecotonal characteristics required to undertake self-purification through biological processes (Lakatos et al 1997, Soler et al. 1991, Chindah *et. al.*, 2007).

It is known that these water bodies have that drain Port Harcourt Municipality played a crucial role in growth and sustaining the development of human communities; however, it is paradoxical that they have undergone degradation in modern times due to various anthropogenic activities (Chindah *et. al.*, 2009). With the attendant increase in population, industrial and commercial activities, untreated municipal, industrial solid wastes and effluents discharged have led to the total degradation of the water quality in many of the stream systems(Ogan, 1988, Ogamba *et. al.*, and Omunakwe *et. al.*, 2009). The consequences include the problem of water pollution rendering water no longer fit for drinking, recreation, as well as for aquatic life. As a result, thousands of children die everyday from diarrhoea and other water, sanitation and hygiene related diseases and many suffer and are weakened by illness (Pandey, 2006). Streams and rivers are vital and vulnerable freshwater systems that are critical for the sustenance of all life. However, the declining quality of the water in these systems threatens their sustainability and is therefore a cause for concern despite their importance in providing various water resources for domestic, industrial, and agricultural purposes (Musaddiq, 2002, Qureshi and Dutka, 1979).

The objective of this research is to find out the pollution level of different streams located in these catchments. The pollution level was determined by examining different physical and chemical parameters of waste water such as Temperature, pH, alkalinity, hardness, dissolved oxygen, BOD, Ammonia, Nitrate, Sulphate and phosphate, and microbiological properties.

2. Materials and methods

2.1 The study area:

Geographically, Port Harcourt is situated at the eastern flank of the New Calabar River, the study streams are   located within the lowland and freshwater swamp zones that accommodate the northern limits of Port Harcourt City and lie between latitude $4^0 43'E - 4^0 50'$ E and longitude $6^0 57'N - 7^0 05'N$. The catchments cover an estimated area of $94.72km^2$ made of flat ground with lithosphere and hydrosphere are interrelated and consequently involved closely related problems including non point source pollutants.

The five streams that drain this catchment and finally empty into Bonny estuary include:-

- Ntawogba Stream
- Miniweja Stream
- Miniokoro Stream
- Minichida Stream and
- Agbonchia Stream.

Ntawogba Stream

Ntawogba stream lies on the extreme west of the municipality and drains the marshy swamp forest up stream of (Rumueme and Rumuepirikom) empties into Amadi creek.

Large ares of the catchment are developed with concrete structures and the lower reach of the stream is concreted.

Miniweja Stream

This stream system drains the Freshwater (Rumuigbo/Rumuola) forest and through various communities and empties into Diobu Creek of Bonny estuary. The greater stretch of the stream channel is degraded while the middle and lower reaches have been more developed with concrete structures.

Miniokoro Stream

Miniokoro stream drains the freshwater swamp forest into Woji creek from where it eventually empties into the Bonny estuary. The entire stream stretch is degraded, built up area with concrete structures and paved roads.

Minichida Stream

Minichida drains the freshwater swamp forest, meanders through communities and empties into Elelenwo creek in Bonny estuary. All the reaches of the stream are degraded by human activities with concrete structures along the water course.

Agbonchia Stream

Agboncha which lies in the east flank of Port Harcourt drains the freshwater swamp forest and empties through Obufe /Elelenwo creek into the Bonny estuary. Development and degradation is fairly low compared with other stream systems but the water body also read recieves effluent of a petrochemical plant.

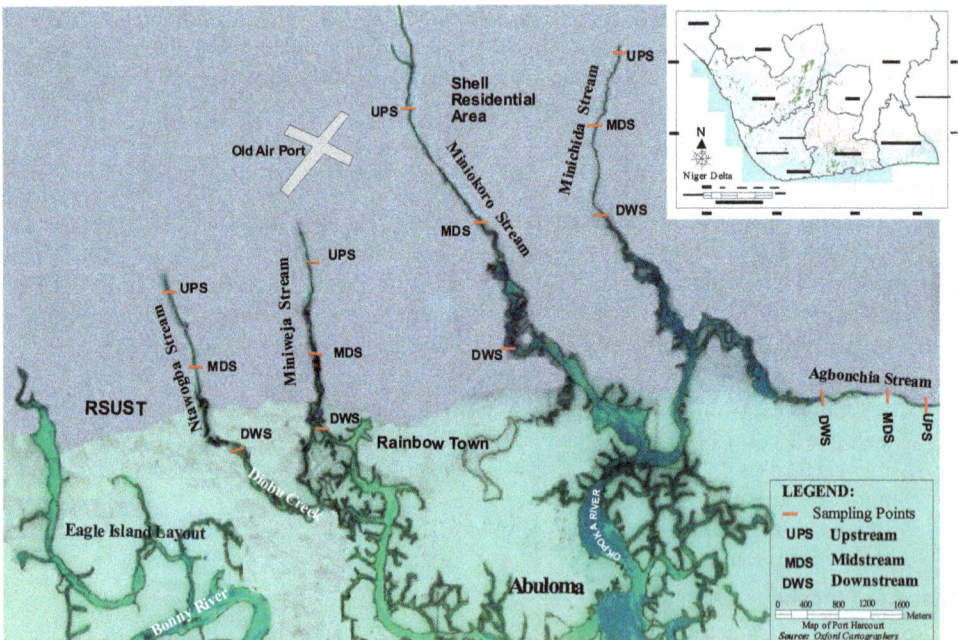

Fig. 1. Map of Niger Delta, Port Harcourt environment showing the 5 study streams

The climatic condition is humid typical of the semi hot equatorial type (Gobo, 1988 and Gobo *et. al.*, 2008). The area experiences heavy rainfall from April to October. Sporadic rainfalls are however experienced during the dry season months of November to March. The mean annual rainfall is estimated to be about 2,405 mm. The prime cause of critical unsanitary conditions of the water bodies is due to the lack of facilities for collection and disposal of waste effectively such that municipal untreated effluent wastewater are discharged into natural surface water drains and sometimes on land and finally through storm water to the stream systems

2.2 Sampling strategy

Samples were collected monthly for 12 months to cover the two main seasons from 6 designated locations from each of the 5 streams (Agboncha stream, Minichida stream, Miniokoro stream, Miniweja stream and Nta-wogba stream). Each of the stream systems, were strategically divided into three segments (upper limit, middle reach and down stream) along the water course considering flow, stretch and human activities. For each segment, two stations were located at an interval of 0.5km. The parameters sampled include temperature, pH, alkalinity, hardness, conductivity, dissolved oxygen, BOD_5, sulphate, ammonia, nitrate, phosphate and microbial properties.

2.3 Physicochemical parameters

In the field, samples for each station were collected with clean 2ml plastic containers at sub-surface level and stored in an ice chest (-4oC). The samples was immediately transferred to the Institute of Pollution Studies (IPS) laboratory for analysis. In the laboratory, analysis was done using procedures as outlined in Standard Methods for the Examination of water and wastewater (1 and 10). Temperature was measured in-situ using a mercury bulb thermometer. pH was measured with a pH meter (Hanna instrument model HI8314). The conductivity was measured using the Horiba water checker model U-10 and Carbon dioxide was measured by the tirimetric method as described in APHA (1998). Dissolved oxygen (DO), and biochemical oxygen demand (BOD_5) were determined using Winkler's method as described in APHA (1998). Other parameters such as ammonia-nitrogen (NH_3-N), nitrate-nitrogen (NO_3-N), sulphate (SO_4^{-2}), and phosphate (PO_4^{-3}) concentrations were determined spectrophotometrically (Spectronic Spectrophotometer 21D), following the procedures as described in APHA (1998). The media used for the bacteriological analysis of water include plate count agar (PCA), nutrient agar (NA), lactose broth (LB), and Eosin Methylene blue agar (EMB). All the media used were weighed-out and prepared according to the manufacture's specification, with respect to the given instructions and directions. A serial dilution method was used for total viable count and the presumptive test for coliforms (APHA,1998).

3. Results

3.1 Minichida

Water temperature was generally high, as expected in the dry season (26.75 ± 1.1 - 27.25 ± 1.27oC) value slightly higher than that of wet season (26.79 ± 0.99 - 26.93 ± 1.32oC). The differences in temperature amongst stations were significant (R^2 = 1) but not significant during the wet season (R^2 = 0.25) Table 1.

Table 1. The mean and standard deviation of the physicochemical and microbial variables for the 3 main ecological limits of the 5 streams for wet and dry seasons

Parameter	Season	Nta-Wogba Upstream	Nta-Wogba Middle Reach	Nta-Wogba Down Stream	Miniweja Upstream	Miniweja Middle Reach	Miniweja Down Stream	Mininkoro Upstream	Mininkoro Middle Reach	Mininkoro Down Stream	MiniChida Upstream	MiniChida Middle Reach	MiniChida Down Stream	Agbonchia Upstream	Agbonchia Middle Reach	Agbonchia Down Stream
$T\,^{\circ}C$	DRY	28.00	28.50	31.00	28.13	30.17	30.58	26.84	28.17	30.33	26.75	27.00	30.33	26.96	26.83	27.17
	SD	1.00	1.05	1.91	0.93	1.60	1.49	1.04	1.94	1.12	1.10	1.26	1.12	1.38	0.41	0.73
	WET	28.07	27.50	28.64	26.72	28.29	28.32	26.22	27.29	29.25	26.86	26.79	26.93	26.43	26.43	26.47
	SD	2.83	1.44	2.63	1.13	2.06	2.49	1.42	1.60	1.40	1.08	0.99	1.32	1.13	1.13	1.12
pH	DRY	6.17	6.19	6.29	6.34	6.55	6.58	6.10	5.90	6.57	6.46	5.97	6.07	6.85	7.00	6.95
	SD	0.03	0.04	0.05	0.35	0.30	0.39	0.38	0.54	0.41	0.30	0.38	0.38	0.45	0.40	0.45
	WET	6.46	6.55	6.57	6.25	6.33	6.37	6.18	6.00	6.35	6.25	6.30	6.19	7.35	7.50	7.30
	SD	0.16	0.13	0.18	0.27	0.38	0.34	0.33	0.41	0.45	0.53	0.46	0.19	0.05	0.10	0.00
CO_2 (mg/l)	DRY	24.42	11.79	12.66	35.10	26.39	24.33	23.90	31.75	18.57	43.55	33.98	35.23	19.73	31.21	29.24
	SD	16.48	4.49	8.90	9.99	4.63	9.02	9.87	12.28	5.50	14.54	8.91	9.57	6.34	10.30	3.18
	WET	36.96	38.10	25.85	40.88	37.67	36.74	34.67	39.67	25.23	32.89	37.96	32.61	22.48	31.96	32.50
	SD	16.20	19.52	11.88	13.37	13.62	17.07	26.63	26.97	6.23	9.51	12.07	10.86	4.65	11.34	9.05
ALKALINITY (mg/l)	DRY	28.10	155.53	60.50	11.84	28.00	32.50	7.17	11.00	31.84	7.17	7.67	7.00	4.50	6.33	7.00
	SD	21.62	144.35	18.06	2.86	13.86	23.65	1.87	2.45	8.31	2.89	1.51	2.43	1.45	1.97	3.05
	WET	21.72	46.29	51.50	12.07	20.57	24.72	7.00	7.43	23.86	7.57	7.71	11.14	4.22	5.86	6.57
	SD	10.24	22.22	19.48	3.22	10.05	10.88	2.56	3.41	10.31	2.33	2.69	4.26	2.10	2.34	2.46
CONDUCTIVITY (μS/cm)	DRY	250.30	1215.00	29566.00	2263.85	13165.00	17190.85	33.34	221.00	18531.67	29.59	26.00	29.00	22.50	39.50	409.25
	SD	271.46	1029.56	13961.91	2633.75	14932.40	16075.35	7.34	211.34	1223.84	5.77	5.55	6.25	8.48	13.90	415.15
	WET	99.14	224.29	13895.72	543.00	5994.30	7888.60	35.72	89.86	1053.57	33.86	28.86	42.29	27.67	37.29	459.00
	SD	38.13	87.53	10074.36	1196.95	11684.30	9742.30	16.22	132.68	1205.89	7.13	8.93	19.96	30.88	9.53	754.54
HARDNESS (mg/l)	DRY	38.50	184.96	2297.60	333.12	1076.80	1438.72	10.88	32.96	1161.20	8.32	6.40	7.52	5.12	5.76	60.80
	SD	25.09	165.11	1128.16	335.97	1150.57	1367.80	19.88	20.73	80.45	5.02	3.36	3.44	2.87	2.97	76.12
	WET	27.02	63.36	1392.00	183.41	781.54	1380.35	19.06	20.30	137.62	12.21	9.34	16.46	4.93	5.21	107.66
	SD	11.52	24.51	848.51	287.88	857.59	1575.43	18.40	17.83	86.91	6.99	5.23	12.43	4.50	3.27	131.78
DO (mg/l)	DRY	1.45	0.44	5.80	3.24	5.97	6.91	1.02	0.93	5.36	1.48	1.10	1.18	2.88	5.04	5.46
	SD	1.59	0.71	4.45	1.01	2.25	3.06	1.02	0.93	1.89	0.71	0.71	0.62	0.94	0.82	1.21
	WET	6.03	4.41	6.41	3.64	6.24	6.44	2.56	4.99	5.12	4.10	3.05	2.19	2.97	4.81	4.90
	SD	7.67	2.62	3.04	1.20	2.11	2.93	0.88	0.94	1.55	2.56	2.62	1.52	0.85	0.74	0.64
BOD (mg/l)	DRY	26.45	34.82	55.25	18.72	23.82	25.56	23.28	29.16	33.85	11.78	18.66	27.76	9.62	14.39	17.32
	SD	9.67	9.43	7.44	5.74	4.84	6.58	3.59	4.17	5.85	6.55	3.41	7.47	0.95	0.97	0.92
	WET	13.45	26.23	37.86	11.65	13.07	14.62	12.92	19.16	22.66	9.52	12.64	16.49	5.75	11.39	19.83
	SD	3.50	8.66	8.54	5.83	5.53	4.67	4.67	2.05	5.63	2.74	2.57	3.13	1.36	2.90	5.90
NH4-N (mg/l)	DRY	0.79	3.39	1.35	0.45	0.19	0.33	0.42	0.31	0.18	0.39	0.59	0.26	0.25	0.20	0.23
	SD	0.89	3.73	1.78	0.42	0.18	0.33	0.50	0.31	0.14	0.40	0.91	0.34	0.31	0.19	0.24
	WET	0.68	1.34	0.82	0.31	0.29	0.38	0.40	0.38	0.38	0.33	0.31	0.38	0.31	0.26	0.27
	SD	0.73	1.49	1.11	0.27	0.21	0.42	0.45	0.30	0.37	0.29	0.30	0.49	0.23	0.20	0.20
NO3-N (mg/l)	DRY	0.67	0.43	0.53	0.81	0.71	0.68	0.55	0.91	0.66	0.56	0.46	0.46	0.53	0.60	0.58
	SD	0.41	0.20	0.28	0.31	0.22	0.18	0.24	0.39	0.28	0.17	0.20	0.17	0.28	0.37	0.25
	WET	0.52	0.55	0.71	0.91	0.61	0.69	0.43	0.35	0.49	0.41	0.35	0.38	0.33	0.45	0.35
	SD	0.14	0.21	0.21	1.33	0.27	0.24	0.17	0.16	0.22	0.21	0.25	0.20	0.19	0.51	0.22
SO4 (mg/l)	DRY	15.50	13.38	275.32	61.81	414.50	603.06	235.12	298.82	466.81	189.67	311.56	299.34	78.69	84.55	210.63
	SD	13.15	6.44	194.26	70.84	359.93	486.01	45.23	12.23	56.41	34.22	57.84	48.93	34.12	41.64	98.57
	WET	9.92	6.29	227.50	18.64	165.04	199.91	300.33	523.45	988.77	342.78	427.28	462.93	85.43	107.36	299.51
	SD	8.74	3.94	124.57	42.17	95.83	54.81	52.18	105.22	96.42	24.78	58.91	95.32	23.78	45.41	68.42
PO4-P (mg/l)	DRY	0.18	0.33	0.17	0.13	0.14	0.15	0.20	0.12	0.14	0.11	0.13	0.11	10.25	14.70	8.80
	SD	0.16	0.35	0.32	0.15	0.13	0.25	0.20	0.18	0.07	0.11	0.06	0.08	8.90	10.00	1.65
	WET	0.32	0.47	0.25	0.08	0.19	0.24	0.38	0.18	0.10	0.09	0.09	0.08	3.90	10.10	60.25
	SD	0.17	0.37	0.25	0.14	0.27	0.24	0.29	0.11	0.06	0.10	0.03	0.20	2.40	2.30	29.35
Total Coliform (cfu/100ml)	DRY	677.00	578.00	986.00	342.00	459.00	533.00	235.12	298.82	466.81	189.67	311.56	299.34	78.69	84.55	210.63
	SD	102.64	87.12	134.00	45.34	101.23	76.80	45.23	12.23	56.41	34.22	57.84	48.93	34.12	41.64	98.57
	WET	452.00	582.00	811.00	621.86	782.15	698.45	300.33	523.45	988.77	342.78	427.28	462.93	85.43	107.36	299.51
	SD	65.00	79.00	122.00	76.33	95.83	54.81	52.18	105.22	96.42	24.78	58.91	95.32	23.78	45.41	68.42
Faecal Coliform (cfu/100ml)	DRY	192.66	225.66	328.67	114.00	153.00	177.54	78.37	99.61	155.60	63.22	103.85	155.60	26.23	28.18	70.21
	SD	22.80	19.36	29.78	10.07	22.49	17.06	10.05	2.71	12.56	7.64	12.83	12.56	7.58	9.25	21.90
	WET	151.67	277.14	195.30	208.63	296.39	262.46	201.45	175.65	197.56	114.85	143.38	197.56	28.66	36.02	100.56
	SD	19.11	23.23	35.88	22.45	28.18	16.12	15.34	30.94	28.35	7.25	17.35	28.35	6.99	13.35	20.12

pH was acidic for both seasons with values slightly more acidic in the dry season (5.97 ± 0.38- 6.46 ± 0.30) than in the wet (6.19 ± 0.19 - 6.30 ± 0.46). Concentrations of alkalinity increased from the upstream to the down stream stations for wet and dry seasons, Seasonal trend demonstrated higher concentrations in the wet season (7.57 + 2.33 - 11.14 ± 4.25mg/l) than during the dry (7.17 ± 2.89 - 8.10 ± 2.43mg/l) season. Spatial differences between the stations in dry season (R^2 = 0.06) were not significant but significant differences were observed during the wet season (R^2 = 0.78) amongst the stations. Carbon dioxide concentrations were relatively higher in the dry season (33.98 ± 8.91 - 43.55 ± 14.54mg/l) than in the wet season (32.61 ± 10.86 - 37.96 ± 12.07mg/l) Table 1.

Conductivity was relatively higher in the wet season (28.86 ± 8.93 - 42.29 ± 8.93µS/cm) than in the dry season (26.0 ± 5.55 - 29.59 ± 5.77 µS/cm) with no clear spatial trend along the water course. Alkaline value was higher in the wet season (7.57 ± 2.33 – 11.14 ± 4.16mg/l) than during the dry season (7.0 ± 2.43 - 7.67 ± 1.51mg/l) with concentrations increasing down stream in the wet but in the dry season no defined sequence was observed and the spatial distributions were not significant in the dry season (R^2 = 0.06) but significance was observed in the wet season (R^2 = 0.78).

Hardness values were higher in the wet season (9.34 + 5.23 - 16.46 + 12.43mg/l) than during the dry season (6.4 ± 3.36 - 8.32 ± 5.02mg/l) and there were no clear spatial patterns between wet and dry seasons Affinity between the stations for wet and dry seasons were not significant but the dry season (R^2 = 0.35) values indicated closer affinity between stations than in the wet season (R^2 = 0.17). Dissolved oxygen concentrations were low in the dry season (1.1 ± 0.71 - 1.48 ± 0.71mg/l) than during the wet season (2.19 ± 1.52 - 4.1 ± 2.58 mg/) with the upstream having the highest dissolved oxygen concentrations than the other limits for both seasons.

BOD5 on the other hand recorded higher concentrations in the dry season (11.78 ± 6.55 - 27.76 ± 7.47mg/l) than in the wet season (9.52 ± 2.74 - 16.49 ± 3.13mg/l).in both seasons, concentrations increased down stream.

Sulphate concentrations tended to decrease down stream for both seasons with wet season (0.85 ± 0.20 - 0.95 ± 0.09mg/l) concentrations being higher than dry season values (0.75 ±0.37 - 0.76 ± 0.38mg/l). However, the distribution of concentrations between the stations indicated closer affinity in the wet season (R^2 = 1) than during the dry season (R^2 = 0.75)

Ammonia nitrogen concentrations were relatively higher in the dry season (0.28 ± 0.34 - 0.59 ± 0.91mg/l) than in the wet season (0.31 ± 0.30 - 0.38 ± 0.49mg/l) and the spatial distribution demonstrated that the middle reach had higher concentrations followed by the upstream and the down stream in that respective order (Table 1).

Nitrate nitrogen concentrations declined downstream with concentrations being relatively higher in the dry season (0.45 ± 0.17 - 0.56 ± 0.17mg/l) than during the wet season (0.35±0.25 - 0.41 ± 0.21mg/l). Also differences between the stations in dry season (R^2 = 0.79) were significant but significance was not observed in the wet season (R^2 = 0.29). Phosphate concentrations were low for both seasons dry season concentrations (0.11± 0.05 - 0.13 ± 0.06mg/l) being slightly higher than wet season concentrations (0.01 ± 0.03 - 0.15 ± 0.08mg/l) Table 1.

The total coliform concentrations were high for both seasons and concentrations consistently increased down stream. Dry season (189.67 ± 34.22 – 289.34 ±48.93 cfu/100ml) concentrations were higher than values observed in the wet season (342.28 ±24.78 – 462.93 ±95.32cfu/100ml). Spatial distribution amongst the three zones indicated significance but with closer affinity between the zones in dry season (R^2 = 0.95) than in the wet season

($R^2 = 0.59$). Correspondingly, the faecal coliform concentrations followed similar seasonal and spatial pattern as observed but concentrations were lower by a magnitude of about 4 times with concentrations for dry season (63.22 ± 7.64 – 103.85 ± 12.83cfu/100ml) being higher than of the wet season (114.85 ± 7.25 – 155.34 ±28.01cfu/100ml) Table 1.

3.2 Agbonchia

Temperature values were high with wet season (26.43 ± 1.13 - 26.47 ± 1.12^{o}C) values not remarkably different from the dry season (26.83 ± 1.38 - 27.17 ± 0.73^{o}C) but spatially distribution amongst the study stations indicated significant difference in wet season ($R^2 = 0.75$) while dry season temperature distributions were not significant ($R^2 = 0.39$) Table 1. In dry season water pH ranged from slightly acidic to neutral while wet season pH was for all the stations, above neutral value. Spatial distributions amongst the stations were significant in dry season ($R^2 = 0.99$) indicating differences in distribution while wet season values were not significant ($R^2 = 0.25$).

Carbon dioxide concentrations were considerably higher in the dry season than in the wet season with values almost increasing down stream for both seasons and differences between the stations were significant for wet ($R^2 = 0.79$) and dry season ($R^2 = 0.60$). Dy season concentrations demonstrated closer affinity than that of wet season (Table 1).

Alkalinity values for both seasons increased down stream and were relatively higher in the dry season (4.50 ± 1.45 - 7.0 ± 3.05 mg/l) than during the wet season (4.22 ± 2.1 - 6.57 ± 2.46mg/l). Spatial differences between stations were positively significant for wet ($R^2 = 0.95$) and dry season ($R^2 = 0.93$). Similarly water hardness increased down stream for both seasons and concentrations were higher in the wet season (4.93 ± 4.50 - 107.66 ± 131.78mg/l) than during the dry season (5.12 ± 2.87 - 60.80 ± 76.12mg/l). The distribution between the stations were significant for wet ($R^2 = 0.75$) and dry ($R^2 = 0.76$) seasons.

Highest conductivity concentrations were observed at the down stream stations which are about 40 - 50 times higher than values observed for the other stations for both seasons. Concentrations for wet season were relatively higher in the wet (27.67 ± 30.88 - 459 ± 755.54μS/cm) than in the dry season (22.50 ± 8.48 - 409 ± 459.15 μS/cm). Spatial differences between the stations was significant in wet season ($R^2 = 0.56$) but not significant in the dry season ($R^2 = 0.20$) Table 1.

Dissolved oxygen concentrations were low and generally increased down stream for both seasons with dry season concentrations generally higher (2.88 ± 0.94 - 5.46± 1.21mg/l) than in the wet (2.97 ± 0.85 - 4.90 ± 0.64mg/l). Spatial differences between the stations for wet ($R^2 = 0.78$) and dry seasons were significant ($R^2 = 0.87$) Table 1.

BOD$_5$ values were considerably high for both wet (5.75 ± 3.77 - 16.83 ± 5.90mg/l) and dry (9.62 ± 0.95 - 17.32 ± 0.90mg/l) seasons. The values consistently season increased down stream in dry season, similarly wet season concentrations at the down stream stations the recorded highest values. However spatial variations between the stations indicated marked differences between the stations for dry ($R^2 = 0.92$) and wet season ($R^2 = 0.69$) Table 1.

Ammonia concentrations were low for both seasons with wet season (0.26 ± 0.20 - 0.31 ± 0.23mg/l) concentrations being higher than in of the dry season (0.20 ± 0.19 - 0.25 ± 0.22mg/l). However spatial distribution of concentrations amongst stations were significant in the wet season ($R^2 = 0.66$) but not significant during the dry season ($R^2 = 0.16$) Table 1. Conversely, nitrate concentrations were relatively higher in the dry season (0.53 ± 0.28 - 0.60 ± 0.23mg/l) than during the wet season (0.33 ± 0.19 - 0.45 ± 0.51mg/l) and difference amongst stations were not significant for wet ($R^2 = 0.01$) and dry season ($R^2 = 0.43$) Table 1.

Sulphate concentrations did not demonstrate any defined spatial distribution pattern within the seasons but wet season concentrations (1.36 ± 0.76 - 57.51 ± 38.72mg/l) were observably higher than that of dry season (1.69 ± 1.58 - 21.90 ± 24.24 mg/l). However, the distribution of concentrations for dry season amongst the stations was significant ($R^2 = 0.89$) but wet season distribution was not significant ($R^2 = 0.01$) Table 1.

Amongst the nutrient variables phosphate had the highest concentrations and values increased down stream especially during the wet season (Table 1). In addition, wet season concentrations (3.9 ± 2.4 - 60.25 ± 59.35 mg/l) were higher than values observed for dry season (8.80 ± 1.65 - 10.25 ± 8.90 mg/l) and the variations amongst the stations for wet ($R^2 = 0.76$) and dry ($R^2 = 0.95$) seasons were significant.

The microbial properties defined by total coliform concentrations were relatively higher in the wet season (85.43 ± 23.78 – 299.51 ± 68.42cfu/100ml) than during the dry season (78.69 ± 34.12 – 210.63 ± 98.57cfu/100ml). The spatial distribution of concentrations amongst the zones for both seasons demonstrated significant positive relationship with the wet season ($R^2 = 0.83$) having closer affinity than the dry season ($R^2 = 0.78$) The faecal coliform concentrations demonstrated similar increasing concentration down stream and concentrations were higher in the wet season (28.66 ± 6.99 – 100.56 ± 20.12 cfu/100ml) than during the dry (26.23 ± 7.58 – 70.21 ± 21.90cfu/100ml) with affinity between zones being significant for both season

3.3 Miniokoro

Temperature values as characteristics of equatorial tropical latitude were high for both dry ($26.84 + 1.04$ - 30.33 ± 1.12°C) and wet (26.22 ± 1.42 - 29.25 ± 1.40°C) seasons with dry season values being relatively higher than in the wet season. The values also increased slightly down stream (Table 1). Regression analysis indicated that dry and wet season distributions between the locations were positively significant with affinity between the stations in the dry ($R^2 - 0.98$) than in the wet ($R^2 = 0.97$). pH was acidic and values were almost uniform for dry (5.9 ± 0.54- 6.57 ± 0.41) and wet (6.0 ± 0.41 - 6.35 ± 0.45) seasons(Table 1). The distribution amongst the stations were not significant for both seasons but dry season values ($R^2 = 0.46$) demonstrated closer affinity between stations than during the wet season ($R^2 = 0.23$).

Carbon dioxide concentration a measure of water acidity was considerably high with values relatively higher in the wet season (25.23 ± 6.23 - 39.67 ± 26.97mg/l) than in the dry season (18.57 ± 5.50 - 31.75 ± 12.28mg/l). The distribution of values amongst the stations was not significant in the dry season ($R^2 = 0.16$) but significant in the wet season ($R^2 = 0.69$) Table 1.

Conductivity values increased consistently down stream for both seasons and dry season (33.34 ± 7.34 - 1831.67 ± 1223.84 µS/cm) values were higher than wet season (35.72 ± 16.22 - 1053.57 ± 1205.89 µS/cm). Similarly alkalinity values increased down stream with dry season (7.17 ± 1.87 - 31.84 ± 8.31mg/l) concentrations being higher than that of wet season (7.0 ± 2.56 - 23.86 ± 10.31mg/l) Table 1.

Chloride concentrations increased down stream by several magnitudes as was observed for alkalinity and conductivity. However, wet season (1.0 ± 0.65 - 314.66 ± 133.93mg/l)concentrations were higher than dry season (1.07 ± 0.74 - 192.48 ± 167.27mg/l) and distribution amongst the stations were similar for wet ($R^2 = 0.76$) and dry ($R^2 = 0.77$)seasons were significant . Hardness concentrations were higher in the dry season (10.88 ± 9.88 - 161.20 ± 80.45mg/l) than in the wet (19.06 ± 18.4 - 137.62 ± 86.91mg/l). The relationship between the stations indicated significance between the stations for both seasons but dry season (R^2 = 0.86) had closer affinity between the stations than in the wet season (R^2 = 0.76)

Dissolved oxygen concentrations were generally high and increased exponentially from upstream to the down stream for dry and wet seasons. Concentrations were slightly higher in the dry season than in the wet season (2.56 ± 0.88 - 5.12 ± 1.55mg/l) and distribution for both dry(R^2 = 0.80) and wet (R^2 = 0.79) seasons demonstrated similar close affinity between station (Table 1).

Biochemical oxygen demand followed a similar sequence of increased concentrations down stream relatively higher concentration being observed in the dry season (23.28 ± 3.59 - 33.85 ± 5.85mg/l)than in the wet (12.92± 4.67 - 22.66 ± 5.63mg/l) Table 1.

Generally nutrient concentrations are low and amongst the nutrient variables only Sulphate demonstrated increasing concentrations from up to down stream. Others such as Phosphate, and Ammonia, had higher concentrations upstream than in other stations. Sulphate had the highest concentrations amongst the nutrient variables with dry season (0.91 ± 0.2 - 45.53 ± 29.30mg/l) concentrations being higher than the wet season (0.92 ± 0.19 - 34.25 ± 21.78mg/l) concentrations and distribution of concentrations amongst the stations for both season were significant (R^2 = 0.75) Table 1.

Nitrate concentrations for dry and wet seasons, were 0.55 ± 0.24 - 0.66 + 0.28mg/l and 0.35 ± 0.16 - 0.49 ± 0.22mg/l respectively. The differences in distribution for wet and dry seasons were not significant with wet season (R^2 = 0.22) demonstrating closer affinity between the stations than the dry season (R^2 = 0.09). Ammonia concentrations were higher in the dry season (0.42 ± 0.5 - 0.91 ± 0.39mg/l) than in wet season (0.35 ± 0.16 - 0.49 ± 0.22mg/l) with the middle reach stations having the highest concentrations for both seasons. The relationship between the stations for wet (R^2 = 0.89) and dry (R^2 = 0.99) seasons where significant with dry season having closer affinity than the wet season. The differences in phosphate concentrations for dry (0.12 ± 0.09 - 0.2 ± 0.26mg/l) and wet season (0.10 ± 0.38 ± 0.29mg/l) seasons were not remarkable but the affinity between the stations were more in the wet season (R^2 = 0.95) than in the dry season (R^2 = 0.50)

As was observed in the other stream systems total coliform concentrations recorded higher counts during the wet season (302.33 ± 52.18 – 588.77 ± 96.42cfu/100ml) than in the dry (235.12 ± 45.23 – 466.81 ± 56.41cfu/100ml) and spatial distribution of concentrations amongst the three zones for both wet (R^2=0.91) and dry(R^2=0.94) seasons were significant(Table 1). The faecal coliform count followed the same increasing concentration pattern down stream in dry season with somewhat different order in the wet season but wet season (201.45± 15.34 – 197.56 ± 28.35cfu/100ml) concentrations being higher than those of dry season (78.37 ± 10.05 – 155.60 ± 12.56 cfu/100ml). In spite of the relative high values recorded in the wet season differences between the zones were not significant (R^2 = 0.02) but dry season distribution were significant(R^2 = 0.94) Table 1.

3.4 Miniweja

Surface water temperatures were high with dry season (28.13±0.98 – 30.58±1.49°C) values being relatively higher than in the wet season (26.72±1.13 -28.29±2.49°C) and temperature tended to increase down stream for both seasons (Table 1). Dry season values (R^2 = 0.87) amongst the stations displayed closer affinity than during the wet season (R^2 = 0.76). pH was slightly acidic for wet(6.25 ± 0.27- 6.37 ± 0.34) and dry (6.24 ± 0.35 - 6.58) seasons and differences between stations were significant with wet season(R^2 = 0.95) demonstrating closer affinity between stations than the dry season (R^2 = 0.80) Table 1. Carbon dioxide concentrations were higher in wet season (36.74 ± 17.07 - 40.88 ± 13.37mg/l) than during the dry season (26.39 + 4.63 - 35.10 + 9.59mg/l) and distribution of concentrations between the stations showed closer affinity in the wet season (R^2 = 0.91) than in the dry season (R^2 = 0.89). Surface water alkalinity generally increased down stream and ranged from 11.84 ± 2.86 - 32.50 ± 23.65mg/l and 12.07 ±3.22 - 24.72 ± 10.88mg/l for dry and wet seasons respectively (Table 1). The relationships between the stations were positively significant with stations in the wet season (R^2 = 0.;96) having closer affinity than in the dry season (R^2 = 0.90). Similarly conductivity values were exceptionally high and increased down stream with higher concentrations occurring during the dry season (2263.85 ± 2433.75 - 17190.85 ± 16075.35µS/cm) than at the wet period (543 ± 1196.95- 7888.60 ± 9742.30µS/cm) Table 1. Affinity between stations was significant for wet (R^2 = 0.93) and dry (R^2 = 0.93) season. Hardness concentrations were high and spatial and seasonal concentrations pattern of increasing values down stream and higher concentrations in the dry season (333.12 ± 335.97 - 1438.72 ± 1367.80mg/l) against the wet season (183.41 ± 287.88 - 1380.35 ± 1575mg/l)as was observed for conductivity. The relationships between the stations for wet (R^2 = 0.99) and dry (R^2 = 0.96) seasons were positively significant.

Dissolved oxygen concentrations for wet and dry seasons were in the ranges of 3.64 ± 1.20 - 6.44± 2.93mg/l and 3.24 ± 1.01 - 6.91 ± 3.01mg/l respectively (Table 1).. Differences between stations were significant with dry season (R^2 = 0.93) having closer affinity than wet season values (R^2 = 0.80). Similarly BOD_5 concentrations increased downstream and concentrations were relatively higher during the dry season (18.72 ± 5.74 - 25.56 + 6.58mg/l) than in the wet season (11.65 ± 5.83 - 14.62 ± 6.67mg/l) Table 1.

High chloride concentrations were observed with relatively higher concentrations in the dry season (446.03 ± 495.13 - 2708.49 ± 2391.26mg/l) than during the wet season (99.15 ± 243.18 - 1380.35 ± 2118.31mg/l) and differences between stations for wet (R^2 = 0.99) and dry (R^2 = 0.97) seasons were significant. Suphate for dry season (61.81 ± 70.84 - 603.01 ± 486.05mg/l) were higher than concentrations in the wet season (18.64 ± 42.17 - 199.91 ± 272.36mg/l) and variations amongst stations for wet (R^2 = 0.89) and dry (R^2 = 0.97) seasons were significant. Ammonia concentrations were relatively higher in the dry season (0.19 ± 0.18 - 0.45 ± 0.42mg/l) than during the wet season (0.29 ± 0.21 - 0.38 ± 0.42mg/l) and variations between stations were only significant in the wet season (R^2 = 0.55) but not significant during the dry season (R^2 = 0.22). Nitrate concentrations appeared relatively higher in the wet season than in the dry and ranged from 0.68 ±0.18 - 0.81 ± 0.31mg/l and 0.61 ± 0.27 - 0.91 ± 1.33mg/l for dry and wet seasons respectively. The affinity between stations were higher in the dry season (R^2 = 0.93) than during the wet season (R^2 = 0.50). Similarly phosphate concentrations spatially tended to increase down stream and wet season concentrations were higher than that of the dry season (0.13 ± 0.12 - 0.15 ± 0.14mg/l),seasonal differences amongst the stations were significant (R^2 = 0.99) for both seasons(Table 1).

Total coliform distributions exhibited obvious seasonal changes (Table 1) with Dry season (342.00 ± 45.34 – 533.00 ± 76.80cfu/100ml) concentrations being relatively lower than wet season concentration (621.86 ± 76.33 – 782.15 ± 95.83cfu/100ml). However the distribution of concentrations amongst the stream course was significant in dry season (R^2 = 0.98) but not significant in wet season (R^2 = 0.98). Faecal coliform recorded lower concentrations against the total coliform with similar seasonal trend such that dry season (114.00 ± 10.07 – 177.54 ± 17.06 cfu/100ml; R^2 = 0.98) concentrations were lower than that of wet season (208.63 ± 22.45 – 296.39 ± 28.18 cfu/100ml; R^2 = 0.37)

3.5 Ntawogba

surface water temperature values were generally high with mean values ranging from 26.83 ± 0.44 -27.08 ± 0.21 in wet season while dry season values ranged from 27.75 ± 0.32o -28.17 ± 0.31oC(Table 1). Spatial variation between stations demonstrated significance for both seasons with affinity between the stations being closer in the wet season (R^2 = 0.96) than during the dry season (R^2 = 0.57).

The pH was slightly acidic for both seasons and differences between the seasons were minimal and values ranged from 6.46 ± 0.16 - 6.57 ± 0.18 and 6.17 ±0.03 - 6.29± 0.05 for wet and dry seasons respectively (Table 1). Spatial differences between the study stations for wet (R^2= 0.10) and dry (R^2=0.10) seasons were not significant. Carbon dioxide concentrations for wet and dry seasons stood at 25.82 ± 11.88 - 38.1 ± 19.52mg/1 and 11.79 ± 4.49 - 24.42 ± 16.48mg/1 and differences amongst the stations were significant demonstrating more affinity in the dry season (R^2 = 0.69) than during the wet season (R^2 = 0.67).

Conductivity values were high, ranging from 188.25 +15.17 - 265.0 ±25µS/cm in the wet season and 251.67 ± 17.69 - 375.08µS/cm in dry season (Table 1). There were relative differences on spatial basis with values increasing down stream and seasonal differences amongst stations were significant with dry season (R^2 = 0.90) demonstrating closer affinity amongst the stations than during the wet season (R^2= 0.90).

Alkalinity values for wet and dry seasons increased down stream with higher concentrations recorded in the dry (62.83 + 13.10 - 89.67 + 16.67mg/1) than during the wet season (10.08 ± 1.76 - 14.00 ± 2.25mg/1) and spatial differences between the stations demonstrated significance for wet (R^2 = 0.96) and dry season (R^2 = 0.97).

There was no clear spatial trend demonstrated in the dissolved oxygen distribution other than the fact that the highest concentrations occurred at the upper limit station for both seasons (Table 1) differences between the stations were significant (R^2 =0.61) while dry season differences between stations were not significant (R^2 = 0.26). In all, concentrations were relatively higher in the wet season (6.50 ± 0.50 - 8.42 ± 0.80 mg/1) than during the dry (5.55 ± 0.48 - 7.35 ± 0.65mg/1). BOD_5 concentrations increased almost exponentially down stream with differences in concentrations between wet and dry seasons being 13.45 ± 3.50 - 37.86 ± 8.54mg/1 and 26.45 ± 9.67 - 55.25 ± 7.44mg/1 respectively. The stations demonstrated similar significant differences for wet (R^2 = 0.98) and dry (R^2 = 0.99) seasons

Ammonia concentrations similarly increased downstream for wet and dry seasons and concentrations were higher in the dry season (0.85±0.14 - 2.10 ± 0.22mg/1) than during the wet season (0.41 ± 0.15 - 0.47± 0.23mg/1) Table 1. Spatially, concentrations between stations were significant during both seasons with stations having closer affinity during the wet season (R^2 = 0.98) than during the dry season (R^2 = 0.57). Sulphate concentrations were in magnitude of about two times higher in the dry (10.40 ± 2.40 - 13.69 ±3.99mg/1) than in the wet season (4.34 ± 1.60 - 5.78 + 1.36mg/1) and concentrations increased down stream during both seasons. Significant differences were observed amongst the stations for both seasons with affinity between stations being observed during the dry season (R^2 = 0.98) than during the wet season (R^2 = 0.53). Nitrate concentrations were comparably high with steady increase in concentration from upstream to down stream station. The differences between stations were significant with closer affinity being observed in the dry season (R^2= 99) than in the wet (R^2= 98). Similarly, phosphate concentrations demonstrated an increasing concentrations from upstream to the downstream limit and differences between stations were significant with closer affinity being observed in the wet season (R^2 = 0.91) than during the dry (R^2 = 0.81). Dry season (0.62 ± 0.09 - 0.99 ± 0.20mg/l) concentrations were higher than that of the wet season (0.41 ± 0.15 - 0.70 ± 0.23mg/l) Table 1.

4. Discussion

Generally, the stream systems maintained high temperature values for both wet and dry seasons and this is a common characteristic reported for the Niger Delta waters (RPI, 1985, NES, 2000) which are located at the equatorial latitude where temperature is consistently high all the year round. In all, a number of associations emerged with temperature such that during the wet season, a strong positive correlation between temperature and Alkalinity (r = 0.69), conductivity (r^2 =0.61), hardness (r =0.60), DO (r^2 =0.73), BOD (r^2 =0.55), So_4 (r^2 =0.61) TC (r^2 =0.76) and FC (r^2 =0.58) Table 2. Similarly, in dry season temperature had significant positive correlation with conductivity (r^2 =0.82), Hardness (r^2 =0.82), DO (r^2 =0.63), BOD (r^2 =0.72), SO4 (r^2 =0.76) Total coliform (r^2 =0.77) and faecal coliform (r^2 =0.78) but negative association was observed for dry season period between temperature and carbon dioxide (r^2 = -0.56) Table 3.

The acidity of a water body is an important factor that determines the suitability of water for various purposes, including toxicity to animals and plants. With the exception of Agbonchia stream whose ph varied from slightly acidic to neutral, the stream systems under study were slightly acidic , showing no consistent spatial and seasonal trends. It is pertinent to observe that while the general values of the water bodies may appear alright comparable to WHO (19 84)limits for potable water the values for such systems in the past had been in the range of 4.5 – 6.0 and 4.8 – 6.5 for wet and dry seasons respectively(NDBDA,1987, Igbinosa and Okoh, 2009). The present pH values are considered high for such soft acid water bodies draining forested wet land with leaf litter that impact humic acid substances that give it the low acidity. The change in pH observed which rather tended toward neutrality might be due to decreased forest floor drainage area, washing of concrete structures during storm and increasing draining of domestic effluent water to the stream.as well as influence of brackish water. pH in the wet season was observed to have significant positive correlation with PO_4 (r^2 =0.58), and negatively correlated with total coliform (r^2 =-0.61) and FC (r^2 =-0.65)Table 2 while in the dry season, pH positively correlated only with PO_4 (r^2 =0.53) and negatively correlated with CO_2 (r^2 =-0.57) Table 3.

Conductivity is a measure of the ability of an aqueous solution to carry an electric current. This ability depends on the presence of ions; on their total concentration, mobility, as well as valence; and the temperature of measurement. The relationship with other parameters of note are the positively correlated with hardness (r^2 =0.97), DO (r^2 =0.65), BOD_5 (r^2 =0.58), NO_3(r^2 =0.55), SO_4 (r^2 =0.96), TC (r^2 =0.69) in the wet season but in the dry season, significant positive associations were observed   between conductivity and DO (r^2 =0.60), BOD_5 (r^2 =0.64), SO_4 (r^2 =0.84), TC (r^2 =0.72) and FC (r^2 =0.72) (Table 2 and 3)

Total hardness of all the water bodies showed higher concentration in the dry season than in the wet season. this is primarily due to reduced inflow and evaporation, while the relative lower concentrations observed may be attributed to increasing inflow and dilution. However to high hardness generally observed in the water bodies may in part be associated the the concrete structure covering the path of the stream. Hardness was found to positively correlation with DO (r^2 =0.67), NO3 (r^2 =0.60), SO4 (r^2 =0.97),TC (r^2 =0.69), and FC (r^2 =0.50) in wet season but in dry season slight variation in the relationships between the attributes such as the positive correlation with DO (r^2 =0.58), BOD (r^2 =0.66), SO4 (r^2 =0.81), TC (r^2 =0.74) and FC (r^2 =0.75) Tables 2 and 3.

	Negatively significant
	Positively significant

Table 2. The correlation coefficient between the physicochemical and biological variables in the wet season

Wet season	T °C	pH	CO₂	ALKALINITY	CONDUCTIVITY	HARDNESS	DO	BOD₅	NH₄-N	NO₃-N	SO₄²⁻	PO₄-P	Total Coliform	Faecal Coliform
T °C	1													
pH	-0.33	1												
CO₂	-0.21	-0.49	1											
ALKALINITY	0.69	-0.14	-0.12	1										
CONDUCTIVITY	0.61	-0.11	-0.22	0.66	1									
HARDNESS	0.60	-0.14	-0.12	0.60	0.97	1								
DO	0.73	0.05	-0.03	0.54	0.65	0.67	1							
BOD₅	0.55	-0.13	-0.21	0.81	0.58	0.47	0.42	1						
NH₄-N	0.31	-0.06	0.07	0.82	0.20	0.14	0.21	0.63	1					
NO₃-N	0.41	-0.24	0.23	0.53	0.55	0.60	0.43	0.26	0.23	1				
SO₄²⁻	0.61	-0.05	-0.15	0.58	0.96	0.97	0.72	0.43	0.09	0.58	1			
PO₄-P	-0.32	0.58	-0.12	-0.25	-0.14	-0.13	0.06	0.07	-0.21	-0.29	-0.15	1		
Total Coliform	0.76	-0.61	0.25	0.70	0.69	0.69	0.56	0.58	0.35	0.68	0.65	-0.32	1	
Faecal Coliform	0.58	-0.65	0.49	0.59	0.44	0.50	0.37	0.38	0.41	0.58	0.45	-0.35	0.87	1

Table 3. The correlation coefficient between the physicochemical and biological variables in the dry season

	T°C	pH	CO₂	ALKALINITY	CONDUCTIVITY	HARDNESS	DO	BOD₅	NH₄-N	NO₃-N	SO₄²⁻	PO₄-P	Total Coliform	Faecal Coliform
T°C	1													
pH	0.25	1												
CO₂	-0.56	-0.57	1											
ALKALINITY	0.40	0.46	-0.67	1										
CONDUCTIVITY	0.82	0.33	-0.43	0.25	1									
HARDNESS	0.82	0.34	-0.44	0.28	1.00	1								
DO	0.63	0.28	-0.19	-0.17	0.60	0.58	1							
BOD₅	0.72	0.17	-0.63	0.56	0.64	0.66	0.22	1						
NH₄-N	0.16	0.38	-0.57	0.95	0.11	0.14	-0.40	0.47	1					
NO₃-N	0.27	-0.31	0.23	-0.30	0.06	0.05	0.52	-0.04	-0.41	1				
SO₄²⁻	0.76	0.18	-0.24	0.13	0.84	0.81	0.65	0.29	-0.08	0.21	1			
PO₄-P	-0.42	0.53	0.02	-0.26	-0.26	-0.26	0.21	-0.49	-0.23	-0.14	-0.24	1		
Total Coliform	0.77	0.26	-0.59	0.57	0.72	0.74	0.18	0.88	0.49	-0.01	0.47	-0.55	1	
Faecal Coliform	0.78	0.29	-0.64	0.67	0.72	0.75	0.16	0.90	0.58	-0.06	0.47	-0.55	0.99	1

Dissolved oxygen is one of the most vital factors in assessing stream quality. Its deficiency directly affects the ecosystem of a stream due to several factors which include physical, chemical, biological and microbiological processes. DO is needed to support biological life in aquatic systems. The levels observed for the study streams are so low that they may not sufficiently support aquatic life including fish. This objectionable low concentration occurred at both seasons, may be associated with the municipal discharges and the attendant organic load and utilization in bacterial decomposition of organic matter. DO in wet season correlated significant with SO_4 (r^2 =0.72), and TC (r^2 =0.56) and in the dry season such associations were observed with NO_3 (r^2 =0.52) and So_4 (r^2 =0.65) Tables 2 and 3.

Biological oxygen demand, being a measure of the oxygen in the water that is required by the aerobic organisms and the biodegradation of organic materials exerts oxygen pressure in the water and increases the biochemical oxygen demand (Abida, 2008). Streams with low BOD_5 have low nutrient levels; and this may account for the general low nutrient status of the stream in most cases.

The increased concentration of BOD_5 implies that oxygen is swiftly depleted in the streams. The consequences of high BOD_5 concentrations are the same as those for low dissolved oxygen: thus organisms are prone to stress, suffocate, and possibly death. In wet season, BOD_5 correlated with NH_4 (r^2 =0.63)and TC (r^2 =0.58) while in dry season the relationships that emerged were significant positive correlation with TC (r^2 =0.88) and Fc (r^2 =0.90) Tables 2 and 3.

Ammonia, a transitional nutrient, generally recorded higher values in the dry season than in the wet season. The distribution of concentration followed a pattern of Nta Wogba > Minchida > ,Minweja > Minikoro > Agboncha in the dry season and in the wet season a slight shift was observed such that the concentration sequence being Nta Wogba > Miniokoro> Minichida > Miniweja > Agboncha

Similarly the same seasonal differences were observed in the distribution of nitrate with higher concentrations in the dry season than in the wet season and the distribution of concentrations being in the decreasing order of Miniweja > miniokoro > Agboncha > Nta wogba > Minichida and Minweja > Ntawogba > Miniokoro >Minichida =Agboncha for dry and wet season periods respectively

The sulphate was the highest of all the nutrients in the different stream and it is considered major composition of seawater following the role of municipal and industrial wastes on sulphate addition to of surface water bodies. The distribution of sulphate concentrations followed a decreasing order of Miniweja stream > Ntawogba stream > Miniokoro stream > Aboncha stream > Minichida stream and Miniweja stream > Ntawogba stream > Agbonchia stream > Miniokoro stream > Minichida stream for dry and wet seasons. However, it is pertinent to note that values observed for Miniweja and Ntawogba were by hundreds of magnitude higher than values observed in the other stream systems

Phosphates as with nitrates are important in assessing the potential biological productivity of surface waters. Increasing concentration of phosphorus and nitrogen compounds in streams or rivers may lead to eutrophication. In this study higher concentrations were recorded in the wet season than in the dry seasons for all the streams and concentrations were considered normal for all the streams except at Agboncha stream in which the distribution of concentration followed a declining order of Agboncha stream > Nta wogba stream > Miniokoro stream >Miniweja stream > Minichida stream and Agboncha stream > Ntawogba stream > Miniokoro stream > Miniweja stream > Minichida stream for dry and wet seasons respectively. The high phosphate value in Agboncha stream may be related in part to Abattoir discharges and petrochemical waste discharges into the system.

The comparison of the variables for the streams using 2 -way Analysis of variance (ANOVA) for the upper limit stations in the wet season demonstrated non significance between the variables (ANOVA = 2.06 , < F (2.08$_{(0.05)}$)) and between streams (ANOVA = 1.88 < F = 2.61$_{(0.05)}$)) Table 4. The middle reach limits of the streams also demonstrated non significance for the variables (ANOVA= 1.15 < F = 2.08$_{(0.05)}$)) and between streams (ANOVA = 1.34 < F = 2.61$_{(0.05)}$)) Table 4. The downstream limits demonstrated a contrary pattern with significance been observed for the variables (ANOVA = 3.06 > F = 2.15$_{(0.05)}$)) but stream differences were also not significant (ANOVA = 1.33 < F = 2.63 $_{(0.05)}$)) Table 4.

Upstream limits					
Source of Variation	*SS*	*df*	*MS*	*F*	*P-value*	*F crit*
Variables	97035.61	10	9703.561	2.06	0.05	2.08
Water bodies	35111.77	4	8777.944	1.879257	0.13	2.61
Error	186838.6	40	4670.966			
Total	318986	54				

Middle Reach limits					
Source of Variation	*SS*	*df*	*MS*	*F*	*P-value*	*F crit*
Variables	7180969	10	718096.9	1.15	0.35	2.08
Streams	3346749	4	836687.2	1.34	0.27	2.61
Error	24964554	40	624113.9			
Total	35492272	54				

Down Stream limits					
Source of Variation	*SS*	*df*	*MS*	*F*	*P-value*	*F crit*
Variables	87980538	9	9775615	3.06	0.01	2.15
Stream	16958067	4	4239517	1.325206	0.28	2.63
Error	1.15E+08	36	3199139			
Total	2.2E+08	49				

Table 4. The 2 way Analysis of variance comparing the variables and the streams at different limits in the wet season

Similar trend was observed in the dry season with differences between variables (ANOVA = 1.38 < F = 2.08 $_{(0.05)}$) and the streams (ANOVA = 1.40 < F = 2.61$_{(0.05)}$) for the upper limit stations were not significant. The middle reach limits also demonstrated same pattern as observed with the upper limit with differences between the variables (ANOVA = 1.30 < F = 2.08 $_{(0.05)}$) and the streams (ANOVA = 1.25 < F = 2.61$_{(0.05)}$) not being significant. The down stream limit demonstrated that the differences between the variable (ANOVA = 2.96 < F = 2.08$_{(0.05)}$) were significant but differences between the streams (ANOVA = 1.24 < F = 2.61$_{(0.05)}$) were not significant (Table 5).

Upstream limits						
Source of Variation	*SS*	*df*	*MS*	*F*	*P-value*	*F crit*
Parameters	1185660	10	118566	1.38	0.23	2.08
Streams	482331.8	4	120582.9	1.40	0.25	2.61
Error	3441178	40	86029.46			
Total	5109170	54				

Middle stream limits						
Source of Variation	*SS*	*df*	*MS*	*F*	*P-value*	*F crit*
Parameters	38261014	10	3826101	1.30	0.27	2.08
Streams	14808576	4	3702144	1.25	0.30	2.61
Error	1.18E+08	40	2950478			
Total	1.71E+08	54				

Down stream limits						
Source of Variation	*SS*	*df*	*MS*	*F*	*P-value*	*F crit*
Parameters	3.63E+08	10	36281955	2.96	0.01	2.08
Streams	60805895	4	15201474	1.24	0.31	2.61
Error	4.91E+08	40	12271158			
Total	9.14E+08	54				

Table 5. The 2 way Analysis of variance comparing the variables and the streams at different limits in the dry season

The five streams have similar physiochemical characteristics apparently because they drain from analogous freshwater systems upstream through the stretch of the city into brackish water systems of the Bonny estuary downstream. The study shows that conductivity values are only higher in dry season in Miniweja out of other streams where the values are generally lower in dry season. The reason could be as a result of the study area of Miniweja being more influenced by brackish water than in any other stream. Minichinda, Nta wogba, Miniokoro and Agboncha streams appear to have more influence of the municipal waste water during wet season.

The similarities in characteristics of the streams are further demonstrated by apparently similar pH values obtained. Naturally, the upstream stations are expected to have much more acidic pH values as a result of vegetation and humic substance released into the forest systems (RPI, 1985, Chindah et. al., 1999, Chindah, 2003, Obunwo, et. al., 2004). Then the pH value increases gradually to become more alkaline as the down stream stations of are approached to the influence of brackish water (RPI, 1985, NDES, 2000, NDDC, 2004 and Izonfuo et. al., 2005). However, in the study, the pH values are apparently uniform with only slight spatial differences indicating that the wastes along the course of the stream have altered the characteristics (Brion and Billen (2000).

Nutrient concentrations are generally low except at the down stream of Miniweja stream where phosphate concentrations were very high. The reason for the general low nutrient concentrationin-spite of the organic load received by the systems may be due to both the high temperature and microbial properties of the water body. Organisms in tropical water bodies are known to quickly use up the nutrients under high temperature condition (Chindah and Braide, 2004 and Chindah et. al., 2005).

This effect is also observed in other parameters. For example, the general low dissolved oxygen concentrations in most streams and the relatively higher values of oxygen recorded in the upstream stations comparative to the mid and down stream stations implies the depletion of oxygen along the water course as it flows down stream. This may suggest that the more waste inputs are received by the streams the more its dissolved oxygen concentration declines. Conversely the BOD_5 values are very high and generally increased down stream. This supports the contention that the increased waste load into the system degrades the water quality as the BOD_5 values far exceed concentrations reported in the baseline studies of some of these streams (NDBDA 1987, and Ogan 1988) Therefore it is our contention that the low oxygen concentrations recorded and the high BOD_5 values for all the streams are strong evidence to suggest the impact of organic load introduced from municipal waste into the streams (Rim-Rukeh et. al., 2007, Hill et. al., 2005 and Chen, 2010). Similarly other indices implicating municipal waste discharges on the stream systems are the high total coliform and faecal coliform concentrations observed in the water bodies which are below concentrations recorded in most of the systems in the past studies (Amadi et. al., 1997, Odokuma and Okpokwasili, 1997 and Ogan 1988). The present total coliform and faecal coliform concentrations indicate the seriousness of the impact of municipal waste water on receiving surface waters and the health hazards implication to ignorant users especially children (Braide et. al., 2004, Okoh et. al., 2005 and 2007). The study shows that the rapid growth of Port Harcourt and associated municipal wastes introduced into the five main steams have caused the deterioration of the water quality of the streams and therefore presents the need for a better waste management system (Chen, 2010).

5. References

Amadi, E. N., Chindah, A. C and Ugoh C. C. (1997). The Effect of Municipal Profainage on The Microflora Of A Black Water Stream In Port Harcourt, Nigeria. Niger Delta Biologia , 2 (1) , 125 – 139.

APHA- American Public Health Association (1998). Standard methods for the examination of water and waste water. 20th ed. APHA-AWWA-WPCF. Washington DC. 1220p.

Braide, S.A., Izonfuo, W.A.L., Adiukwu,P.U., Chindah, A.C., and Obunwo, C.C. (2004).Water quality of Miniweja stream, a swamp forest stream receiving non point source waste   discharges in eastern Niger Delta, Nigeria. Scientia Africana, 3(1) 1-8.

Brion, N. and Billen, G. (2000). Wastewater as a source of nitrifying bacteria in river systems: the case of the River Seine downstream from Paris. Wat. Res. 34, (12):. 3213-3221,

Chen G., Cao X. Song C. and Zhou Y. (2010). Adverse Effects of Ammonia on Nitrification Process: the Case of Chinese Shallow Freshwater Lakes. Water Air Soil Pollut. 210:297-306.

Chindah, A. C; Hart A. I. and Atuzie B (1999). A preliminary investigation on the effects of municipal waste discharge on the macrofauna associated with macrophytes in a small fresh water stream in Nigeria. Afri. J of Applied Zool. 2, 29 - 33.

Chindah, A.C. (2003). The physico-chemistry phytoplankton and periphyton of a swamp forest streams in the lower Niger Delta. Scientia Africana 2(1&2) 106-116.

Chindah, A.C. and Braide, S. A (2004). The physicochemical quality and phytoplankton community of tropical waters: A case of 4 biotopes in the lower Bonny River, Niger Delta, Nigeria. Caderno de Pesquisa . Ser. Bio. Santa Cruz do Sul Vol 16 (2), 7-37.

Chindah, A. C., Braide, S. A. and Izundu, E. (2005). Treatment Of Municipal Wastewater Quality Using Sunlight. Caderno de Pesquisa . Ser. Bio. Santa Cruz do Sul Vol 17 (2), 27-45.

Chindah, A.C., Braide, S.A., Amakiri, J., Izundu, E. (2007).Succession of phytoplankton in a municipal waste water treatment system under sunlight. Revista Cientifica UDO Agricola 7(1): 258-273.

Chindah, A.C., Braide S.A, Amakiri J.and Ajibule O.O.K. (2009). Periphyton Succession in a waste water treatment pond. Revista UDO Agricola 9(3): 672 – 680.

Gobo, A.E.(1988) Relationship Between Rainfall Trends And Flooding In The Niger -Benue River Basins J. Meteorol U.K.; 13 (132): 220-24.

Gobo, A.E., Ubong, I.U., Ede, P.N . (2008). Rainfall intensity analysis as a tool for hydrological and agricultural practises in Southern Nigeria. The internat J. Meterol.; 33(334): 343-50.

Hill, D.D., Owens, W.E., and Tchounwou, P.B.C. (2005). Comparative assessment of the physico-chemical and bacteriological qualities of selected streams in Louisiana. Int J Environ Res Public Health. 2(1):94-100.

Igbinosa, E. O. and Okoh, A. I. (2009) Impact of discharge wastewater effluents on the physico-chemical qualities of a receiving watershed in a typical rural community. Int. J. Environ. Sci. Tech., 6 (2), 175-182.

Izonfuo, W.A.L., Chindah, A. C., Braide, S.A., and Lawson D. A.(2005). Physicochemical Characteristics of Different Ecotonal Streams In A Rapidly Developing Metropolis In The Niger Delta, Nigeria . Caderno de Pesquisa . Ser. Bio. Santa Cruz do Sul Vol 17 (2), 91-105.

Lakatos G.; M. K. Kiss, M. Kiss and P. Juha'sz. (1997). Application of constructed wetlands for wastewater treatment in Hungary. Wat. Res. 15 (5): 341-346. laundary detergents. Chemosphere 17: 2175-2182.

Musaddiq, M.(2002). Surface water quality of Morna river at Akolaa. Pollut. Res.,19(4), 685-691.

NDBDA- Niger Delta Basin Development Authority (1987). The Chemical composition of Niger Delta Rivers. Final report on the Environmental Pollution Monitoring of the Niger Delta basin of Nigeria. Vol 5.i-xvii, 1-145.

NDDC- Niger Delta Development Commission (2004).Biodiversity of the Niger Delta Environment Niger Delta Development commission master plan Project Final report.

NDES- Niger Delta Environmental Survey (2000). Ecological zonation and habitat classification. 2nd Phase Report 2, Vol.1: 1-66.

Obunwo, C.C., Braide, S.A., Izonfuo,W.A.L., and Chindah, A.C (2004) Influence of urban activities on the water quality of fresh water streams in the Niger Delta, Nigeria. Journal of Nigerian Environmental Society (JNES) 2: (2) 196-209.

Odokuma , l. O. and Okpokwasili, G. C. (1997). Seasonal influences of the organic pollution monitoring of the New Calabar River, Nigeria Environmental Monitoring and Assessment 45, (1), 43-56.

Ogan, M T (1988). Examination of surface waters used as sources of supply in the Port Harcourt area. II. Chemical hydrology. Archiv fuer Hydrobiologie, Supplement. Vol. 79, no. 203, pp. 325-342.

Ogamba, E.N. Chindah, A.C., Ekweozor, I.K.E., Onwuteaka, J. N (2004). Water quality and phytoplankton in Elechi creek complex of the Niger Delta. *Journal of Nigerian Environmental Society* (JNES) 2: (2) 121-130.

Okoh, A. I.; Barkare, M. K.; Okoh, O. O.; Odjadjare, E., (2005). The cultural microbial and chemical qualities of some waters used for drinking and domestic purpose in a typical rural setting of Southern Nigeria. J. Appl. Sci., 5 (6), 1041- 1048.

Okoh, A. I.; Odjadjare, E. E.; Igbinosa, E. O.; Osode, A.N., (2007). Wastewater treatment plants as a source of microbial pathogens in the receiving watershed. Afr. J. Biotech. 6 (25), 2932-2944.

Onderka, M., Pekarova, P., Miklanek, P. Halmova, D. and Pekar, J. (2010). Examination of the Dissolved Inorganic Nitrogen Budget in Three Experimental Microbasins with Contrasting Land Cover—A Mass Balance Approach. Water Air Soil Pollut . 210:221–230.

Onuoha, G.C., Chindah, A.C. Oladosu, G.A. and Ayinla O.A.. (1991). Effect of organic fertilization on pond productivity and water quality of fish ponds at Aluu, Nigeria. NIOMR 74, 1 - 12.

Pandey, R. (2007). Some effects of untreated wastewater of Bombay (India) on Brassica and Spinacea oleracea. Agr. Res., 12(2): 34-41.

Pekarova, P., Miklanek, P., Onderka, M., & Kohnova, S. (2009). Water balance comparison of two small experimental basins with different vegetation cover. Biologia, 64 (3), 487–491.

Qureshi, A. A. and Dutka, B. J. (1979). Microbiological studies on the quality of urban storm water runoff in Southern Ontario, Canada. Water Research, 13, (10), 977-985.

Rafiu A. O., Roelien D. P. and Isaac R. (2007). Influence of discharged effluent on the quality of surface water utilized for agricultural purposes African Journal of Biotechnology. 6 (19), 2251-2258.

Rim-Rukeh, A. , Ikhifa, G. O. and Okokoyo, P. A. (2007). Physico-Chemical Characteristics of Some Waters Used for Drinking and Domestic Purposes in the Niger Delta, Nigeria. Environmental Monitoring and Assessment. 128 (1-3), 475-482,

RPI –Research Planning Institute (1985).. Environmental Baseline Studies for the establishment of Control Criteria and Standards against Petroleum Related Industries in Nigeria, RPI- Columbia South Carolina, USA. RPI/R/84/4/15-1

Sheikh, K.H. and Irshad, M. (1980). Wastewater effluents from a tannery: Their effects on soil and vegetation in Pakistan. Envi. Conser., 7(4): 319-324.

Soler, A.; Saez, J., Llorens, M.,. Martinez, I., Torrella, F., and. Berna, L. (1991). Changes in physico-chemical parameters and photosynthetic microorganisms in a deep wastewater self-depuration lagoon. Wat. Res.,25 (6): 689-695.

World Health Organization (WHO), (1984). Guidelines Wastewater. 19th Edn. American water Works for Drinking Water Quality. Health Criteria and other Association, Water Environment Federation. Supporting Information, WHO, Geneva, Vol: 1.

Wahid, A., S.S. Ahmad, M.G.A. Nasir. (1999). Water pollution and its impact on fauna and flora of a polluted stream of Lahore. Acta Scient., 9(2): 65-74.

8

Wastewater Management

Peace Amoatey (Mrs) and Professor Richard Bani
Department of Agricultural Engineering,
Faculty of Engineering Sciences,
University of Ghana,
Ghana

1. Introduction

Wastewater is water whose physical, chemical or biological properties have been changed as a result of the introduction of certain substances which render it unsafe for some purposes such as drinking. The day to day activities of man is mainly water dependent and therefore discharge 'waste' into water. Some of the substances include body wastes (faeces and urine), hair shampoo, hair, food scraps, fat, laundry powder, fabric conditioners, toilet paper, chemicals, detergent, household cleaners, dirt, micro-organisms (germs) which can make people ill and damage the environment. It is known that much of water supplied ends up as wastewater which makes its treatment very important. Wastewater treatment is the process and technology that is used to remove most of the contaminants that are found in wastewater to ensure a sound environment and good public health. Wastewater Management therefore means handling wastewater to protect the environment to ensure public health, economic, social and political soundness (Metcalf and Eddy, 1991).

1.1 History of wastewater treatment

Wastewater treatment is a fairly new practice although drainage systems were built long before the nineteenth century. Before this time, "night soil" was placed in buckets along streets and workers emptied them into "honeywagon" tanks. This was sent to rural areas and disposed off over agricultural lands. In the nineteenth century, flush toilets led to an increase in the volume of waste for these agricultural lands. Due to this transporting challenge, cities began to use drainage and storm sewers to convey wastewater into waterbodies against the recommendation of Edwin Chadwick in 1842 that "rain to the river and sewage to the soil". The discharge of waste into water courses led to gross pollution and health problems for downstream users.

In 1842, an English engineer named Lindley built the first "modern" sewerage system for wastewater carriage in Hamburg, Germany. The improvement of the Lindley system is basically in improved materials and the inclusion of manholes and sewer appurtenances — the Lindley principles are still upheld today. Treatment of wastewater became apparent only after the assimilative capacity of the waterbodies was exceeded and health problems became intolerable. Between the late 1800s and early 1900s, various options were tried until in 1920, the processes we have today were tried. Its design was however empirical until midcentury. Centralized wastewater systems were designed and encouraged. The cost of wastewater treatment is borne by communities discharging into the plant.

Today there have been great advances to make portable water from wastewater. In recent times, regardless of the capacity of the receiving stream, a minimum treatment level is required before discharge permits are granted (Peavy, Rowe and Tchobanoglous, 1985). Also presently, the focus is shifting from centralized systems to more sustainable decentralized wastewater treatment (DEWATS) especially for developing countries like Ghana where wastewater infrastructure is poor and conventional methods are difficult to manage (Adu-Ahyia and Anku, 2010).

1.2 Objectives of wastewater treatment

Wastewater treatment is very necessary for the above-mentioned reasons. It is more vital for the:

Reduction of biodegradable organic substances in the environment: organic substances such as carbon, nitrogen, phosphorus, sulphur in organic matter needs to be broken down by oxidation into gases which is either released or remains in solution.

Reduction of nutrient concentration in the environment: nutrients such as nitrogen and phosphorous from wastewater in the environment enrich water bodies or render it eutrophic leading to the growth of algae and other aquatic plants. These plants deplete oxygen in water bodies and this hampers aquatic life.

Elimination of pathogens: organisms that cause disease in plants, animals and humans are called pathogens. They are also known as micro-organisms because they are very small to be seen with the naked eye. Examples of micro-organisms include bacteria (e.g. *vibro cholerae*), viruses (e.g. enterovirus, hepatits A & E virus), fungi (e.g. *candida albicans*), protozoa (e.g *entamoeba hystolitica, giardia lamblia*) and helminthes (e.g. *schistosoma mansoni, asaris lumbricoides*). These micro-organisms are excreted in large quantities in faeces of infected animals and humans (Awuah and Amankwaa-Kuffuor, 2002).

Recycling and Reuse of water: Water is a scarce and finite resource which is often taken for granted. In the last half of the 20th century, population has increased resulting in pressure on the already scarce water resources. Urbanization has also changed the agrarian nature of many areas. Population increase means more food has to be cultivated for the growing population and agriculture as we know is by far the largest user of available water which means that economic growth is placing new demands on available water supplies. The temporal and spatial distribution of water is also a major challenge with groundwater resources being overdrawn (National Academy, 2005). It is for these reasons that recycling and reuse is crucial for sustainability.

1.3 Types of wastewater

Wastewater can be described as in the figure below.

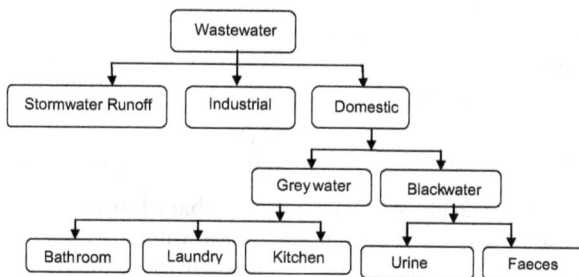

Fig. 1. Types of Wastewater

2. Definition of concepts and terminology

Stormwater Runoff is water from streets, open yard etc after a rainfall event which run through drains or sewers.

Industrial wastewater is liquid waste from industrial establishments such as factories, production units etc.

Domestic wastewater also known as municipal wastewater is basically wastewater from residences (homes), business buildings (e.g. hotels) and institutions (e.g. university). It can be categorized into greywater and blackwater.

Greywater also known as sullage is liquid waste from washrooms, laundries, kitchens which does not contain human or animal excreta.

Blackwater is wastewater generated in toilets. Blackwater may also contain some flush water besides urine and faeces (excreta). Urine and faeces together is sometimes referred to as night soil.

Sewage is the term used for blackwater if it ends up in a sewerage system.

Septage is the term used for blackwater if it ends up in a septic tank.

Sewerage system is the arrangement of pipes laid for conveying sewage.

Influent is wastewater which is yet to enter in a wastewater treatment plant or liquid waste that is yet to undergo a unit process or operation.

Effluent is the liquid stream which is discharged from a wastewater treatment plant or discharge from a unit process or operation.

Sludge is the semi-solid slurry from a wastewater treatment plant.

On-Site System: this is wastewater disposal method which takes place at the point of waste production like within individual houses without transportation. On- site methods include dry methods (pit latrines, composting toilets), water saving methods (pour-flush latrine and aqua privy with soakage pits and methods with high water rise (flush toilet with septic tanks and soakage pit, which are not emptied).

Off-Site System: in this system, wastewater is transported to a place either than the point of production. Off- site methods are bucket latrines, pour-flush toilets with vault and tanker removal and conventional sewerage system.

Conventional sewerage systems can be combined sewers (where wastewater is carried with storm water) or separated sewers.

Septic Tank is an on-site system designed to hold blackwater for sufficiently long period to allow sedimentation. It is usually a water tight single storey tank.

Faecal sludge refers to all sludge collected and transported from on-site sanitation systems by vacuum trucks for disposal or treatment.

Unit Operation: this involves removal of contaminants by physical forces.

Unit Process: this involves biological and/or chemical removal of contaminants.

Wastewater Treatment Plant is a plant with a series of designed unit operations and processes that aims at reducing certain constituents of wastewater to acceptable levels.

3. Characteristics of wastewater

Depending on its source, wastewater has peculiar characteristics. Industrial wastewater with characteristics of municipal or domestic wastewater can be discharged together. Industrial wastewater may require some pretreatment if it has to be discharged with domestic wastewater. The characteristics of wastewater vary from industry to industry and therefore would have different treatment processes — for example a cocoa processing company may have a skimming tank in its preliminary treatment stage to handle for instance spilt cocoa butter while a beverage plant may skip this in the design. In general, the contaminants in wastewater are categorized into physical, chemical and biological. Some indicator measured to ascertain these contaminants include (Peavy, Rowe and Tchobanoglous, 1985 & Obuobie et al., 2006):

Physical
- **Electrical Conductivity** (EC) indicates the salt content
- **Total Dissolved Solids** (TDS) comprise inorganic salts and small amounts of organic matter dissolved in water
- **Suspended solids** (SS) comprises solid particles suspended (but not dissolved)in water

Chemical
- **Dissolved Oxygen** (DO) indicates the amount of oxygen in water
- **Biochemical oxygen demand** (BOD) indicates the amount of oxygen required by aerobic microorganisms to decompose the organic matter in a sample of water in a defined time period.
- **Chemical oxygen demand** (COD) indicates the oxygen equivalent of the organic matter content of a sample that is susceptible to oxidation by a strong chemical oxidant
- **Total Organic Compound** (TOC)
- **NH4-N** and **NO3-N** show dissolved nitrogen (Ammonium and Nitrate, respectively).
- **Total Kjeldhal Nitrogen** is a measurement of organically-bound ammonia nitrogen.
- **Total-P** reflects the amount of all forms of phosphorous in a sample.

Biological
- **Total coliforms** (TC) is encompassing faecal coliforms as well as common soil microorganisms, and is a broad indicator of possible water contamination.
- **Faecal coliforms** (FC) is an indicator of water contamination with faecal matter. The common lead indicator is the bacteria *Escherichia coli* or *E. coli*.
- **Helminth** analysis looks for worm eggs in the water

3.1 Process of wastewater treatment

Due to the nature of contaminants in wastewater — physical, chemical and biological, the unit operations and processes in wastewater treatment can also be categorized as such. The units operations and processes in Waste-water treatment are summarized as follows (Economic and Social Commission for Western Asia (ESCWA), 2003):

Physical unit operations
- Screening
- Comminution
- Flow equalization
- Sedimentation
- Flotation
- Granular-medium filtration

Chemical unit operations
- Chemical precipitation
- Adsorption
- Disinfection
- Dechlorination

- Other chemical applications

Biological unit operations
- Activated sludge process
- Aerated lagoon
- Trickling filters
- Rotating biological contactors
- Pond stabilization
- Anaerobic digestion

3.2 Levels of wastewater treatment

There are three broad levels of treatment: primary, secondary and tertiary. Sometimes, preliminary treatment precedes primary treatment.

Preliminary treatment: removes coarse suspended and grits. These can be removed by screening, and grit chambers respectively. This enhances the operation and maintenance of subsequent treatment units. Flow measurement devices, often standing-wave flumes, are necessary at this treatment stage (FAO, 2006).

Primary treatment removes settleable organic and inorganic solids by sedimentation and floating materials (scum) by skimming. Up to 50% of BOD5, 70% of suspended solids and 65% of grease and oil can be removed at this stage. Some organic nitrogen, organic phosphorus, and heavy metals are also removed. Colloidal and dissolved constituents are however not removed at this stage. The effluent from primary sedimentation units is referred to as primary effluent (FAO, 2006).

Secondary treatment is the further treatment of primary effluent to remove residual organics and suspended solids. Also biodegradable dissolved and colloidal organic matter is removed using aerobic biological treatment processes. The removal of organic matter is when nitrogen compounds and phosphorus compounds and pathogenic microorganisms are removed. The treatment can be done mechanically like in trickling filters, activated sludge methods rotating biological contactors (RBC) or non-mechanically like in anaerobic treatment, oxidation ditches, stabilization ponds etc.

Tertiary treatment or advance treatment is employed when specific wastewater constituents which cannot be removed by secondary treatment must be removed. Advance treatment removes significant amounts of nitrogen, phosphorus, heavy metals, biodegradable organics, bacteria and viruses. Two methods can be used effectively to filter secondary effluent — traditional sand (or similar media) filter and the newer membrane materials. Some filters have been improved, and both filters and membranes also remove helminths. The latest method is disk filtration which utilizes large disks of cloth media attached to rotating drums for filtration (FAO, 2006).

At this stage, disinfection by the injection of Chlorine, Ozone and Ultra Violet (UV) irradiation can be done to make water meet current international standards for agricultural and urban re-use.

4. Methods of wastewater treatment

There are conventional and non-conventional wastewater treatment methods which have been proven and found to be efficient in the treatment of wastewater. Conventional methods compared to non-conventional wastewater treatment methods has a relatively high

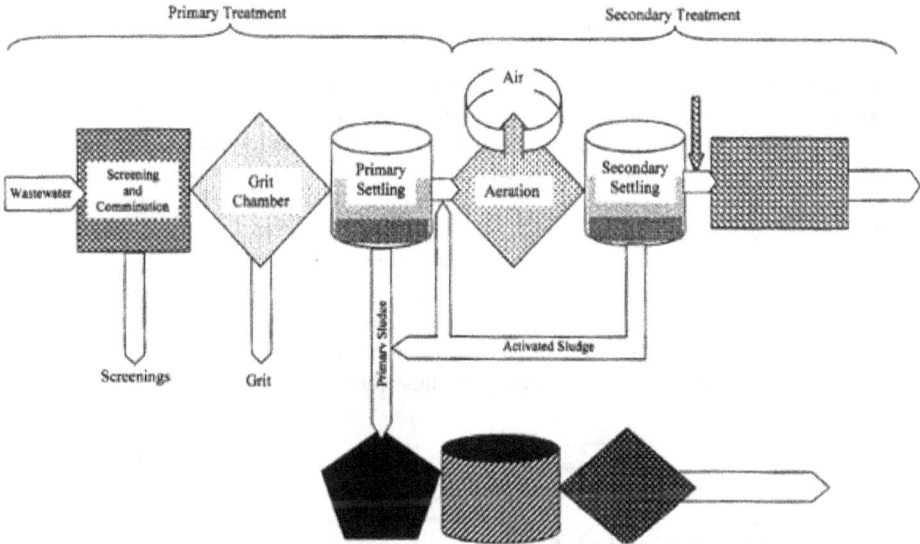

Source: NPTEL (accessed 2010)

Fig. 2. Typical Wastewater Treatment Plant

level of automation. Usually have pumping and power requirements. They require skilled labour for operation and maintenance of the system

4.1 Conventional methods

Examples of conventional wastewater treatment methods include activated sludge, trickling filter, rotating biological contactor methods. Trickling filters and Rotating Biological Contactors are temperature sensitive, remove less BOD, and trickling filters cost more to build than activated sludge systems. Activated sludge systems are much more expensive to operate because energy is needed to run pumps and blowers (National Programme on Technology Enhanced Learning (NPTEL), 2010).

These methods are discussed in detail in the subsequent sections.

4.1.1 Activated sludge

Activated sludge refers to biological treatment processes that use a suspended growth of organisms to remove BOD and suspended solids. It is based on the principle that intense wastewater aeration to forms flocs of bacteria (activated sludge), which degrade organic matter and be separated by sedimentation. The system consists of aeration and settling tanks with other appurtenances such as return and waste pumps, mixers and blowers for aeration and a flow measurement device. To maintain the concentration of active bacteria in the tank, part of the activated sludge is recycled.

Primary effluent (or plant influent) is mixed with return activated sludge to form mixed liquor which is aerated for a specified length of time. By aerating the system, activated sludge organisms use the available organic matter as food, thereby, producing stable solids and more organisms. The suspended solids produced by the process and the additional organisms become part of the activated sludge. The solids are then separated from the wastewater in the settling tank and are returned to the influent of the aeration tank (return activated sludge). Periodically the excess solids and organisms are removed from the system (waste activated sludge) to enhance the performance of the system.

Factors such as temperature, return rates, amount of oxygen available, amount of organic matter available, pH, waste rates, aeration time, and wastewater toxicity affect the performance of an activated sludge treatment system. A balance therefore must be maintained between the amount of food (organic matter), organisms (activated sludge) and dissolved oxygen (NPTEL, 2010).

Activated Sludge systems are requires less space compared to trickling filter and has high effluent quality. The disadvantage is that BOD is higher at one end of the tank than the other the microorganisms will be physiologically more active at that end than the other unless a complet mixing activated sludge system process is used. Presently there are 11 activated sludge plants in Ghana, mainly installed by the large hotels (Obuobie, et al., 2006).

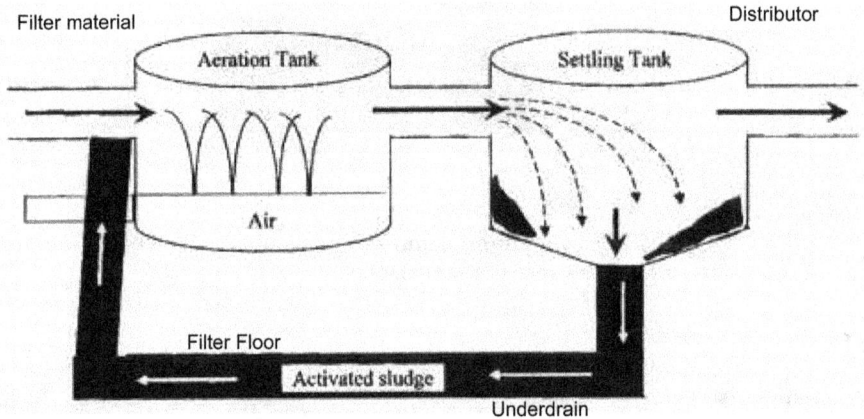

Source: Mountain Empire College, 2010

Fig. 3. An activated Sludge System

4.1.2 Trickling filter:

It is a growth process in which microorganisms responsible for treatment are attached to an inert packing material. It is made up of a round tank filled with a carrier material (volcanic rock, gravel or synthetic material). Wastewater is supplied from above and trickles through filter media allowing organic material in the wastewater to be adsorbed by a population of microorganisms (aerobic, anaerobic, and facultative bacteria; fungi; algae; and protozoa) attached to the medium as a biological film or slime layer (approximately 0.1 to 0.2 mm thick).

Degradation of organic material by the aerobic microorganisms in the outer part of the slime layer occurs. As the layer thickens through microbial growth, oxygen cannot penetrate the medium face, and anaerobic organisms develop. The biological film continues to grow to such a point that microorganisms near the surface cannot cling to the medium, and a portion of the slime layer falls off the filter. This process is known as sloughing. The sloughed solids are picked up by the underdrain system and transported to a clarifier for removal from the wastewater (US EPA, 2000).

Trickling filters are efficient in that effluent quality in terms of BOD and suspended solids removal is high. Its operational costs are relatively low due to low electricity requirements. The process is simpler compared to activated sludge process or some package treatment plants. Its operation and maintenance requirements is however high due to the use of electrical power. Skilled labour is required to keep the trickling filter running trouble-free: e.g. prevent clogging, ensure adequate flushing, control filter flies. It is suitable for some relatively wealthy, densely populated areas which have a sewerage system and centralized wastewater treatment; also suitable for greywater treatment.

It also requires more space compared to some other technologies and has potential for odour and filter flies (NPTEL, 2010).

This method has been widely used in Ghana. There are 14 trickling filter plants in Accra though they have broken down.

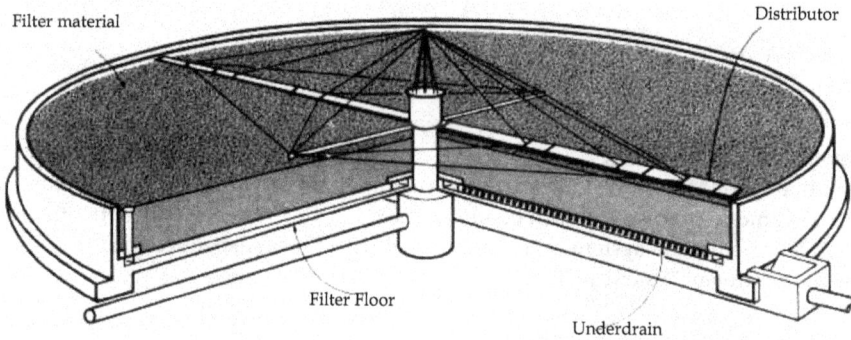

Source: ESCWA, 2003

Fig. 4. Cross section of a trickling filter

4.1.3 Rotating biological contactors

Rotating biological contactors (RBCs) consist of vertically arranged, plastic media on a horizontal, rotating shaft. The plastics range from 2 – 4 m in diameter and up to 10 mm thick (Peavy, Rowe ad Tchobanoglous, 1985). The biomass-coated media are alternately exposed to wastewater and atmospheric oxygen as the shaft slowly rotates at 1–1.5 rpm (necessary to provide hydraulic shear for sloughing and to maintain turbulence to keep solid in suspension), with about 40% of the media submerged. High surface area allows a large, stable biomass population to develop, with excess growth continuously and automatically shed and removed in a downstream clarifier. Thichness of biofilm may reach 2 – 4 mm depending on the strength of wastewater and the rotational speed of the disk.

RBC systems are relatively new, though it appeared to be best suited to treat municipal wastewater (Peavy, Rowe ad Tchobanoglous, 1985), they have been installed in many petroleum facilities because of their ability to quickly recover from upset conditions (Schultz, 2005). The RBC system is easily expandable should the need arise, and RBCs are also very easy to enclose should volatile organic content containment become necessary. RBCs have relatively low power requirements and can even be powered by compressed air which can also aerate the system. They follow simple operating procedures and thus require a moderately skilled labour. RBCs are however capital intensive to install and sensitive to temperature.

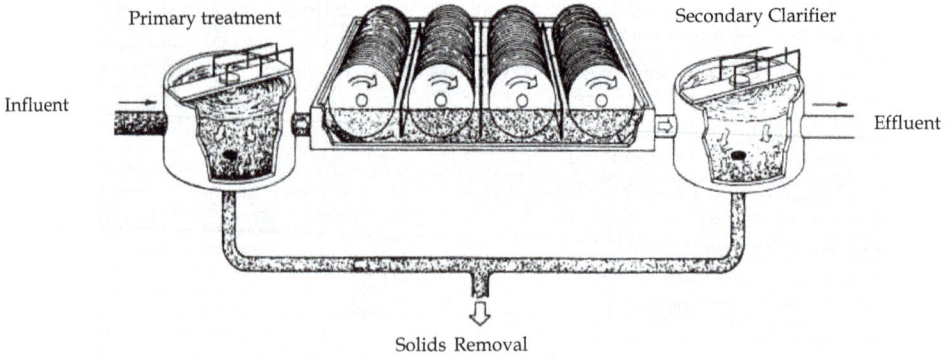

Source: ESCWA, 2003

Fig. 5. Rotating Biological Contactors

4.1.4 Membrane bioreactors

This method performs more than just one treatment step. Membrane bioreactor (MBR) systems are unique processes, which combine anoxic- and aerobic-biological treatment with an integrated membrane system that can be used with most suspended-growth, biological wastewater-treatment systems.

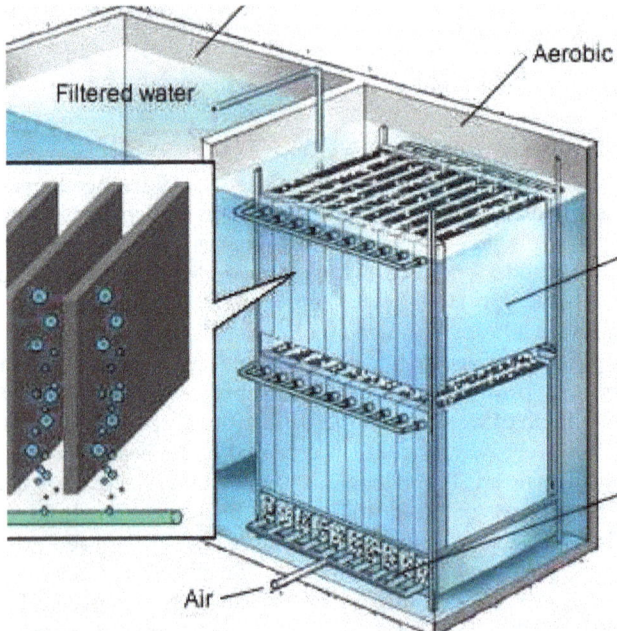

Source: Google Images

Fig. 6. Membrane Bioreator

Wastewater is screened before entering the biological treatment tank. Aeration within the aerobic-reactor zone provides oxygen for biological respiration and maintains solids in suspension. MBR relies on submerged membranes to retain active biomass in the process. This allows the biological process to operate at longer than normal sludge ages (typically 20-100 days for a MBR) and to increase mixed-liquor, suspended-solids (MLSS) concentrations (typically 8,000-15,000 mg/l) for more effective removal of pollutants. High MLSS concentrations reduce biological-volume requirements and the associated space needed to only 20–30% of conventional biological processes.

MBRs cover a small land area as it eliminates the need for secondary clarifiers, which equates to a huge savings in both footprint and concrete costs. They can operate at higher biomass concentrations (MLSS) than conventional treatment processes. Facility can be expanded by simply adding more membranes to existing basins without expanding land cover. For reuse quality, it does not require tertiary treatment, polymer addition, or any further treatment processes to meet standards. This reduction in the number of unit processes further improves system reliability and reduces operation activities (TEC, 2010). The generally high effluent quality reduces the burden on disinfection in the treatment process.

4.2 Non-conventional methods

These are low-cost, low-technology, less sophisticated in operation and maintenance biological treatment systems for municipal wastewater. Although these systems are land intensive by comparison with the conventional high-rate biological processes, they are often more effective in removing pathogens and do so reliably and continuously if system is properly designed and not overloaded (FAO, 2006). Some of the non-conventional methods include stabilization ponds, constructed wetlands, oxidation ditch, soil aquifer treatment.

4.2.1 Waste stabilization ponds

Waste Stabilization Ponds are man-made, shallow basins which comprise of a single series or several series of anaerobic, facultative or maturation ponds. This is a low-technology treatment process with 4 or 5 ponds of different depths with different biological activities. Treatment of the wastewater occurs as constituents are removed by sedimentation or transformed by biological and chemical processes (National Academy, 2005).The anaerobic ponds are mainly designed for the settling and removal of suspended solids as well as the breakdown of some organic matter (BOD_5). In facultative ponds, organic matter is further broken down to carbon dioxide, nitrogen and phosphorous by using oxygen produced by algae in the pond. Maturation ponds usually remove nutrients and pathogenic micro-organisms, thus primary treatment occurs in anaerobic ponds while secondary and tertiary treatment occurs in facultative and maturation ponds respectively (Awuah, 2002).

Anaerobic ponds are usually between 2-5 m deep and receive high organic loads equivalent to 100g BOD_5 and m^3/d leading to anaerobic conditions throughout the pond (Mara et $al.$, 1992). If properly designed, anaerobic ponds can remove 60% of BOD_5 at 200 C.

Facultative ponds are 1-2 m deep and usually receive the effluent from an anaerobic pond. In some designs, they receive raw wastewater acting as primary facultative pond. In facultative ponds organic loads are lower and allows for algal growth which accounts for the dark green colour of wastewater. Algae and aerobic bacteria generate oxygen which breaks down BOD_5. Good wind velocity generates mixing of wastewater in ponds thus leading to uniform mixing of BOD_5, oxygen, bacteria and algae which better stabilizes waste.

Maturation ponds are usually shallow ponds of about 1.0-1.5 m deep allowing aerobic conditions in for the treatment of facultative pond effluents. Further reduction of organic matter, nutrients and pathogenic microorganisms occurs here. Algal population in maturation ponds is more diverse and removal of nitrogen and ammonia is more prominent.

In Ghana so far, stabilization ponds have worked very well due to the convenient climatic conditions. It usually flows under gravity from one pond to the other and mostly does not require any pumping. It is less energy dependent thus plant activities cannot be interrupted due to power cuts. Its disadvantages however include odour problems and it requires a large area of land to function properly. Presently there are 21 stabilisation ponds in Ghana mainly in Accra and Kumasi. Some of them like the Tema Community 3, Achimota, have been closed Various combinations and arrangement of ponds are possible. The figure below shows some possible combinations.

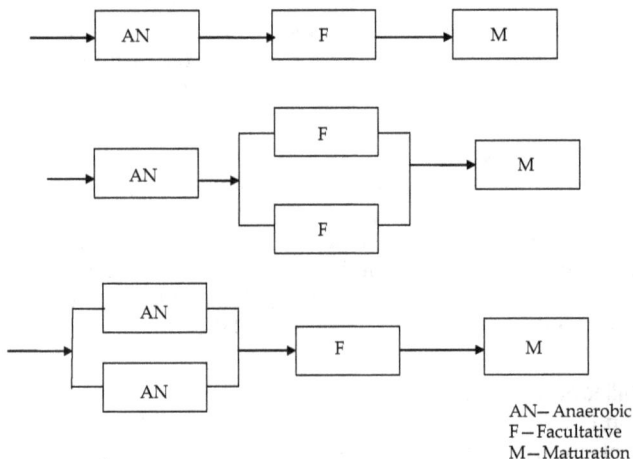

AN— Anaerobic
F — Facultative
M — Maturation

Fig. 7. Various Arrangement of Waste Stabilisation Ponds

4.2.2 Constructed wetlands

Constructed Wetlands (CW's) are planned systems which are designed and constructed to employ wetland vegetation to assist in treating wastewater in a more controlled environment than occurs in natural wetlands (Kayombo et al., 2000). They are an eco-friendly and a suitable alternative for secondary and tertiary treatment of municipal and industrial wastewater. They are suitable for the removal of organic materials, suspended solids, nutrients, pathogens, heavy metals and toxic pollutants. They are not ideal for the treatment of raw sewage, pre-treatment of industrial wastewater to maintain the biological balance of the wetland ecosystem.

There are two types of CW's namely Free Water Surface (FWS) and Subsurface Flow (SSF) systems. As the name suggests, with FWS, water flows above the ground and plants are rooted in the sediment layer below the water column. With SSF, water flows through a porous media such as gravels in which the plants are rooted. From a public health perspective, SSF should be used in primary treatment of wastewater because there is no direct contact of wastewater with atmosphere.

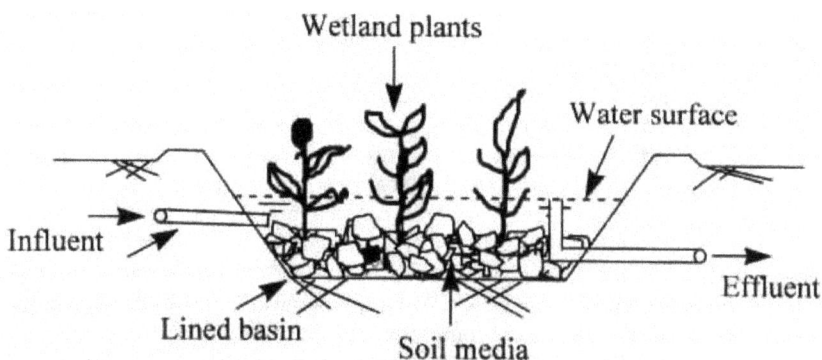

Source: ESCWA, 2003

Fig. 8. Free Water Surface System

The SSF is mostly anoxic or anaerobic as oxygen supplied by the roots of plants is used up in biofilm growth and as such does not reach the water colomn. The flow of water in SSF can be horizontal or vertical (Kayombo *et al.*, 2000). FWS are suitable for treating secondary and tertiary effluents and also providing habitat due to aerobic conditions at and near the surface of the water. There condition at the bottom sediment is however anoxic.

Wetlands plants or macrophytes utilized in CW's include Cattails (*Typha latifolia sp*), *Scirpus* (Bulrus), *Lemna* (duckweed), *Eichornia crassipes* (water hyacinth), *Pistia stratiotes* (water lettuce) *Hydrocotyle* spp. (pennywort), *Phragmites* (reed) have been known and used in constructed wetlands.

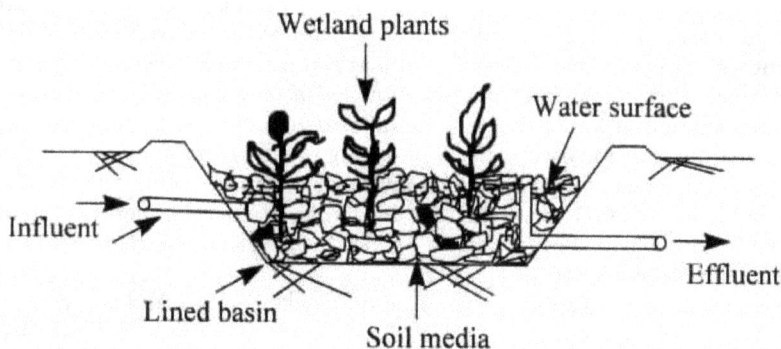

Source: ESCWA, 2003

Fig. 9. Sub-surface flow system

CW's are relatively cheaper to construct operate and easy to maintain. This is an important decision variable for developing countries. In Egypt, according to Hendy (2006), between 2000 and 2004, a 60 acre artificial wetland constructed cost 25% the cost of conventional sewage treatment plant.

They provide effective and reliable treatment of wastewater and are tolerant to fluctuating hydrologic and contaminant loading rates. With the example in Egypt, $9 million (US) was spent to treat an initial volume of 25,000 metric tons per day. After a year of use, it was determined that the wetland was capable of treating 40,000 metric tons per day (Hendy,

2006). Also a study conducted by Ratnapriya *et al.,* (2009) revealed over 60% removal of BOD5, COD, nitrogen among others.

CW's also provide indirect benefits such as enjoying the scenic views of green spaces, encouraging wildlife habitats and providing recreational and educational centres. Again, in Egypt, the fishing industry is expanding since the wastewater was no longer being discharged directly into the waterways, the local fisheries improved. According to Hendy (2006), nitrates and heavy metals were filtered out, leaving the fish healthier, larger and in abundant quantity. This indirectly led to poverty reduction.

They however have some disadvantages such as land requirements, its design and operation criteria is presently imprecise. CW's are biologically and hydrologically complex and its process dynamics are not completely understood. Sometimes there are cost implications of gravels fills and site grading during construction (Kayombo *et al.,* 2000). It must be emphasized that if properly designed, constructed wetlands should not breed pests and mosquitoes.

In Ghana, there are not many CW's. There is presently a pilot SSF donor CW in Tema. This plant is not entirely low-cost as it was designed with some energy dependent units.

4.2.3 Oxidation ditches

An oxidation ditch is a modified activated sludge biological treatment process that utilizes hydraulic retention time of 24 - 48 hours, and a sludge age of 12 - 20 days. to remove biodegradable organics. Oxidation ditches are typically complete mix systems, but can be modified. Typical oxidation ditch treatment systems consist of a single or multichannel configuration within a ring, or oval. Preliminary treatment, such as bar screens and grit removal, normally precedes the oxidation ditch. Primary settling prior to an oxidation ditch is sometimes practiced and tertiary filters may be required after clarification, depending on the effluent requirements. Disinfection is required and reaeration may be necessary prior to final discharge. Horizontally or vertically mounted aerators provide circulation, oxygen transfer, and aeration in the ditch. Flow to the oxidation ditch is aerated and mixed with return sludge from a secondary clarifier. The mixing process entrains oxygen into the mixed liquor to foster microbial growth and the motive velocity ensures contact of microorganisms with the influent. Aeration increases dissolved oxygen concentration but decreases as biomass takes up oxygen during mixing in the ditch. Solids also remain in suspension during circulation (USEPA, 2000).

They require more power than waste stabilization ponds less land, and are easier to control than processes such as activated sludge process. A typical process flow diagram of treatment plant using an oxidation ditch is shown in Figure 10.

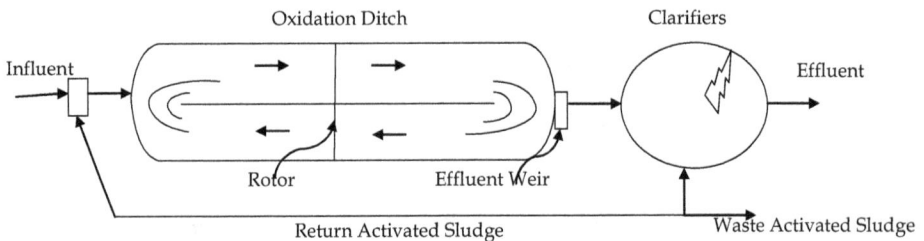

Fig. 10. Oxidation Ditch

4.2.4 Upflow anaerobic sludge blanket (UASB)

Upflow anaerobic sludge blanket is an anaerobic process using blanket of bacteria (see Figure 11) to absorb polluting load. It is a form of anaerobic digester which forms a blanket of granular sludge which suspends in the tank. Wastewater flows upwards through the blanket and is processed (degraded) by the anaerobic microorganisms. The upward flow combined with the settling action of gravity suspends the blanket with the aid of flocculants. Small sludge granules begin to form whose surface area is covered in agregations of bacteria. In the absence of any support matrix, the flow conditions create a selective environment in which only those microorganisms, capable of attaching to each other, survive and proliferate.

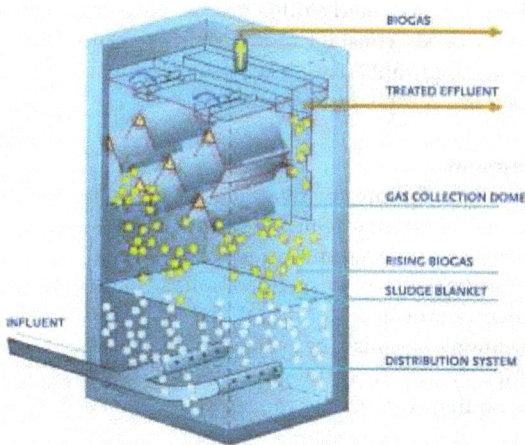

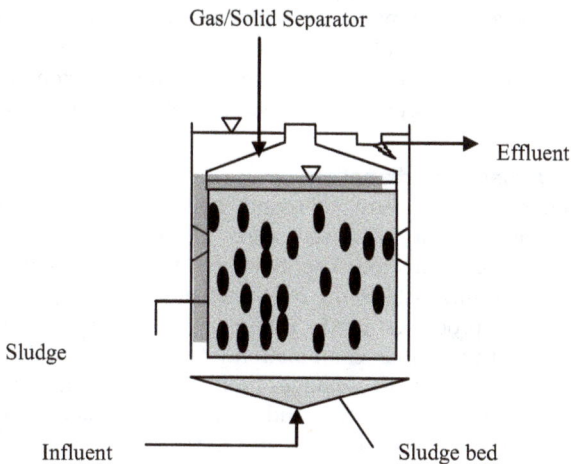

Source: Google Images

Fig. 11. Upflow Anaerobic Sludge Blanket

Eventually the aggregates form into dense compact biofilms referred to as granules. The UASB reactor works best when desirable micro-organisms are retained as highly active and fast settling granules. In the UASB reactor, when high solids retention time is met, separation of gas, sludge solids from the liquid occurs. The special Gas-Solid-Liquid Separators in the reactor enable collection of biogas and recycle of anaerobic biomass. Biogas contains 50 to 80% methane.

UASB is suitable for the primary treatment of high-COD mainly soluble industrial effluents. It can also be used for the treatment of wastewater effluents of low and medium strength. It is suited to hot climates Low energy requirement, less operation and maintenance, lower skill requirement for operation, less sludge production, resource recovery through biogas generation and stabilized waste as manure. UASBs however have relatively poor effluent quality than processes such as activated sludge process (Tare and Nema, 2010).

The technology however, needs constant monitoring to ensure that the sludge blanket is maintained, and not washed out. The heat produced as a by-product of electricity generation can be reused to heat the digestion tanks.

4.2.5 Soil aquifer treatment

Soil matrix has quite a high capacity for treatment of normal domestic sewage, as long as capacity is not exceeded. Partially-treated sewage effluent is allowed to infiltrate in controlled conditions to the soil. The unsaturated or "vadose" zone then acts as a natural filter and can remove essentially all suspended solids, biodegradable materials, bacteria, viruses, and other microorganisms. Significant reductions in nitrogen, phosphorus, and heavy metals concentrations can also be achieved. After the sewage, treated in passage through the vadose zone, has reached the groundwater it is usually allowed to flow some distance through the aquifer for further purification before it is collected through the aquifer.

Soil-aquifer treatment is a low-technology, advanced wastewater treatment system. It also has an aesthetic advantage over conventionally treated sewage since effluent from an SAT systems is clear and odour-free and it is viewed as groundwater either than effluent. Discharge effluent should travel sufficient distance through the system and residence times should be long enough, to produce effluent of desired quality (FAO, 2006).

4.3 Faecal sludge treatment and disposal

Sewage sludge contains organic and inorganic solids that were found in the raw wastewater. Sludge from primary and secondary clarifier as well as from secondary biological treatment need to be treated. The generated sludge is usually in the form of a liquid or semisolid, containing 0.25 to 12 per cent solids by weight, depending on the treatment operations and processes used. Sludge is treated by means of a variety of processes that can be used in various combinations. Thickening, conditioning, dewatering and drying are primarily used to remove moisture from sludge, while digestion, composting, incineration, wet-air oxidation and vertical tube reactors are used to treat or stabilize the organic material in the sludge (ESCWA, 2003).

Thickening: Thickening is done to increase the solids content of sludge by the reduction of the liquid content. An increase in solids content from 3 to about 6 per cent can decrease total sludge volume significantly by 50 per cent. Sludge thickening methods are usually physical in nature: they include gravity settling, flotation, centrifugation and gravity belts.

Stabilization: Sludge stabilization is aimed at reducing the pathogen content, eliminate offensive odours, and reduce or eliminate the potential for putrefaction. Some methods used for sludge stabilization include lime stabilization, heat treatment, anaerobic digestion, aerobic digestion and composting (ESCWA, 2003).

5. Wastewater reuse in agriculture

Irrigation with wastewater is both disposal and utilization and indeed is an effective form of wastewater disposal (as in slow-rate land treatment). However, some degree of treatment must normally be provided to raw municipal wastewater before it can be used for agricultural or landscape irrigation or for aquaculture.

In many industrialized countries, primary treatment is the minimum level of preapplication treatment required for wastewater irrigation. It may be considered sufficient treatment if the wastewater is used to irrigate crops that are not consumed by humans or to irrigate orchards, vineyards, and some processed food crops (FAO, 2006).

Nutrients in municipal wastewater and treated effluents are a particular advantage as supplemental fertilizers. Success in using treated wastewater for crop production will largely depend on adopting appropriate strategies aimed at optimizing crop yields and quality, maintaining soil productivity and safeguarding the environment. Several alternatives are available and a combination of these alternatives will offer an optimum solution for a given set of conditions. The user should have prior information on effluent supply and its quality. Wastewater effluent can be blended with conventional water or solely used. Heavy metal concentrations in streams used for irrigation in and around urban centres such as Accra and Kumasi have been sometimes found to be beyond recommended levels for irrigation purposed and should therefore may pose a health concern.

Countries must develop standards in congruence with the WHO guidelines and enforce it.

6. Industrial wastewater treatment

In general, the type of plant to be installed depends on the characteristics of the wastewater produced from that industry. The basic principle according to Kamala and Kanth Rao (1989) however is waste prevention by good housekeeping practices that will ultimately result in volume reduction and strength reduction. Industrial wastewater is treated the same way as domestic or municipal sewage — preliminary, primary, secondary and advanced treatment levels. Most of the treatment methods discussed is also applicable. There could however be peculiarities with different industrial depending on their major contaminant e.g. heavy metals, dye, etc.

Industrial wastewater in Ghana is generated from breweries, distilleries, textile, chemical & pharmaceuticals and institutions and hotels which are mainly situated in Accra and Tema. In the Western and middle belt of Ghana, mining activities are predominant and the major polluter of our rivers. EPA-Ghana grants permits to industries and requires industries to install or build an in-house waste treatment plant. EPA-Ghana takes samples quarterly from these industrial wastewater plants for testing in their own laboratories for monitoring purposes. Most of those who have permit have treatment plants though not all of them are functioning properly.

In recent years, there has been a growth of small-scale industries in the fruits and food processing industries in the Tema, light industrial area which do not have the resources to build treatment plant. Most of these small-scale industries empty their wastewater into nearby drains without treatment. In Kumasi, the principal generators of industrial wastewater in Kumasi are the two breweries, a soft drink bottling plant and an Abattoir.

7. Status of wastewater treatment plants in Ghana

The use of on-site treatment systems is quite extensive. Individual and community/residential based septic tanks are the most preferred. Septic tanks only partially treat sewage, and the effluent is still rich in organic material. The septic tank has to be emptied from time to time and the disposal of the septic sludge causes severe public health and environmental particularly in urban areas.

Major wastewater treatment methods found in Ghana includes stabilisation ponds, trickling filters and activated sludge plants. According to a recent survey, there are 46 wastewater treatment plants in Ghana. More than half of all treatment plants in Ghana are in the Greater Accra region, mainly in the capital city of Accra and port city of Tema. Brong Ahafo and Upper West regions have no treatment plants at all. The stabilization pond method is the most extensively used with almost all faecal sludge and large-capacity sewage treatment plants using the method. Most trickling filters and activated sludge plants recorded have a low capacity and belong to private enterprises like larger hotels.

Only about 10 of the treatment plants are operational (Obuobie *et al.*, 2006) and it is not clear if these plants meet the EPA effluent guidelines. This can be attributed to the fact that the conventional methods are energy dependent and also when the mechanical parts become faulty, the part has to be imported making it too expensive to maintain. Low-cost, low-technology methods are however manageable.

8. Challenges of wastewater management

Wastewater management though not technically difficult can sometimes be faced with socio-economic challenges. A few of the challenges are discussed below.

8.1 Infrastructure

Most often than not, wastewater infrastructure are not the priority of most politicians and therefore very little investment are made. It is however important to consider wastewater infrastructure as equally important as water treatment plant because almost all the water produced ends up as wastewater.

8.2 Pollution of water sources

Effects of wastewater effluent on receiving water quality is enormous, it changes the aquatic environment thus interrupts with the aquatic ecosystem. The food we eat contains carbonaceous matter, nutrients, trace elements and salts and are contained in urine and faeces (black water).

Medications (drugs), chemicals and in recent times hormones (contraceptives) are also discharged into the wastewater treatment plant. Discharge guidelines must be strictly adhered to. This will ensure sustainability of water sources for posterity.

The precautionary and the polluter-pays principles which prevent or reduce pollution to the wastewater have proven to be very efficient in the industrialized countries and should be adapted in developing countries as well.

8.3 Choice of appropriate technology

Because the economy of most developing countries is donor driven, funds for wastewater plants are mainly from donors. For this reason, they tend to propose the technology which should be adopted. For this reason, when the beneficiaries, take over the facility, its management of the operations and maintenance of parts become quite challenging as the technical expertise, power requirements etc are not sustainable.

8.4 Sludge production

Treatment of wastewater results in the production of sewage sludge. There must be a reliable disposal method. If it must be used in agriculture, then the risks involved must be taken into consideration. Due to the presence of heavy metals in wastewater, it is sometimes feared that agricultural use may lead to accumulation of heavy metals in soils thereby contaminating of yields.

8.5 Reuse

Effluents which meet discharge standards could be used for agricultural purposes such as aquaculture or for irrigation of farmlands. The challenge however is that if wastewater treatment plants are not managed and continuously monitored to ensure good effluent quality, reuse becomes risky.

9. Conclusion

Wastewater is and will always be with us because we cannot survive without water. When water supplied is used for the numerous human activities, it becomes contaminated or its characteristics is changed and therefore become wastewater. Wastewater can and must be treated to ensure a safe environment and foster public health. There are conventional and non-conventional methods of wastewater treatment and the choice of a particular method should be based on factors such as characteristics of wastewater whether it from a municipality or industry (chemical, textile, pharmaceutical etc.), technical expertise for operation and maintenance, cost implications, power requirements among others.

In most developing countries like Ghana, low-cost, low-technology methods such as waste stabilization ponds have been successful whilst conventional methods like trickling filters and activated sludge systems have broken down. Effluent which meets set discharge standards can be appropriately used for aquaculture and also irrigation. Though there are a few challenges in waste water management, they can be surmounted if attention and the necessary financial support is given to it.

10. References

Adu-Ahyiah, M. and Anku, R. E. "Small Scale Wastewater Treatment in Ghana (a Scenerio)" Retrieved, 03-10-2010:1-6

Awuah, E. and Amankwaa-Kuffuor, R., (2002) "Characterisation of Wastewater, its sources and its Environmental Effects" I-Learning Seminar on Urban Wastewater Management

Economic and Social Commission for Western Asia (2003) "Waste-Water Treatment Technologies: A General Review" United Nation Publication

Food and Agricultural Organisation (2006) Wastewater Treatment http://www.fao.org/docrep/t0551e/t0551e06.htm#TopOfPage

Hendy, S. M. H. (2006) "Wastewater Management and Reuse in Egypt" Regional Workshop on Health Aspects of Wastewater Reuse in Agriculture Amman, Jordan 30 October –3 November 2006

National Programme on Technology Enhanced Learning (2010) "Wastewater Treatment" Course   Notes www.nptel.iitm.ac.in/courses/Webcourse-contents/IIT accessed 01-09-2010

Kamala, A. and Kanth Rao, D. L., (1989) Environmental Engineering Tata MacGraw-Hill Publishing 121-218

Kayombo S., Mbwette T.S.A. Katima J.H. Y, Ladegaard N., and Jorgensen S.E. (2000) "Waste Stabilisation Ponds and Constructed Wetlands Design Manual UNEP-IETC and DANIDA Publication.

Keraita B. N. and Dreschsel, P (2004) "Agricultural Use of Untreated Wastewater in Ghana' International Research Development Center Publication 11-19

Metcalf and Eddy, Inc. (1991) "Wastewater Engineering": Treatment Disposal and Reuse, third edition. New York: McGraw-Hill.

Metcalf and Eddy, Inc. (2003) "Wastewater Engineering: Treatment and Reuse," Fourth edition.: McGraw-Hill, New York.

Mountain Emire Community College (2010) "Activated Sludge" Lecture Notes http://water.me.vccs.edu/courses/env110/Lesson17_print.htm accessed 03-10-2010.

National Academy (2005) "Water Conservation, Reuse, and Recycling": Proceedings of an Iranian-American Workshop National Academies Press, Washington

Niyonzima, S. Awuah, E. and Anakwa, A. O. (2008) Grey Water Treatment using Constructed Wetland in Ghana. Abstract—Switch Urban Water 2008

Obuobie, E., Keraita B. N., Danso, G., Amoah, P., Cofie, O. O. Raschid-Sally, L. and Dreschsel, P (2006) "Sanitation and Urban Wastewater Management" Book Chapter in Irrigated Urban Vegetable Farming in Ghana: Characteristics, Benefits and Risks http://www.cityfarmer.org/GhanaIrrigateVegis.html

Peavy, S. H., Rowe, D. R. and Tchobanoglous, G., (1985) Environmental Engineering, International Edition MacGraw-Hill 207-322.

Ratnapriya, E. A. S. K., Mowjood, M I M, De Silva, R P, and Dayawansa, N D K (2009) "Evaluation of constructed wetlands for efficiency of municipal solid waste leachate treatment in Sri Lanka" Presentation at DAAD_GAWN Alumni Expert Seminar, March 2010.

Schultz, T. E. (2005) "Biotreating Process Wastewater: Airing the Options, Chemical Engineering.

Tare, V. And Nema, A. "UASB Technology: Expectations and Reality" Retrieved 01-10-2010

Treatment Equipment Company "An Introduction to Membrane Bioreactor Technology"
    Company Information Sheet retrieved 01-10-2010
US-EPA, (2000) "Tricking Filters" Wastewater Technology Fact Sheet Washington DC.
US-EPA, (2000) "Oxidation Ditches" Wastewater Technology Fact Sheet Washington DC.

9

Wastewater from Table Olive Industries

G.M. Cappelletti, G.M. Nicoletti and C. Russo
Dipartimento SEAGMeG – University of Foggia,
Italy

1. Introduction

In several Mediterranean countries the production of table olives plays an important part in the national economy. Moreover, in recent years there has been a worldwide increase in the production and consumption of these olives (Figure 1). From 2003 to 2009 the major olive-producing countries were: Spain (with an average of 503,300 tonnes per annum, representing approximately 26% of world production), Egypt (with an average of 299,600 tonnes per annum – 15.4% of world production) and Turkey (with an average of 230,800 tonnes per annum – 11.9% of world production). The major olive-consuming countries were: the USA; with an average of 217,600 tonnes per annum – 11.2% of world consumption), Spain, with an average of 197,700 tonnes per annum – 10.1% of world consumption) and Turkey (with an average of 183,700 tonnes per annum – 9.4% of world consumption).

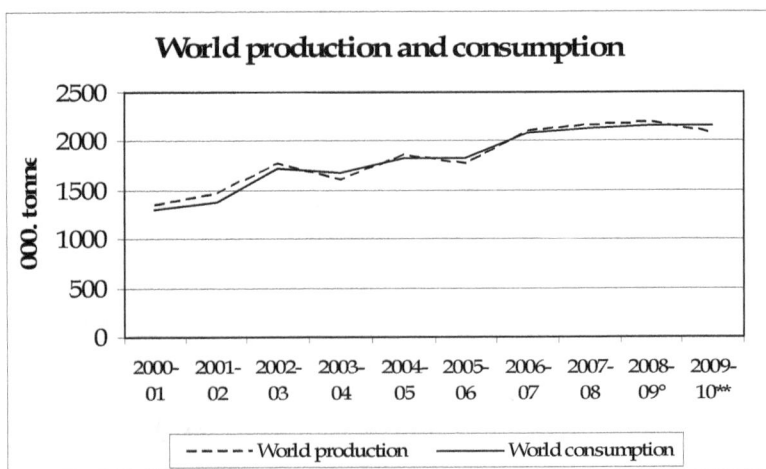

Source: www.internationaloliveoil.org

° provisional data   **expected data

Fig. 1. Table olives world production and consumption

During the same period the major olive-exporting countries were: Spain (exporting an average of 181,700 tonnes per annum), Argentina (exporting an average of 65,000 tonnes per annum), Morocco (exporting an average of 63,500 tonnes per annum) and Egypt (exporting an average of 62,800 tonnes per annum). The countries which imported the most olives were: the USA (which imported an average of 129,600 tonnes per annum), Brazil (which imported an average of 61,400 tonnes per annum) and Russia (which imported an average of 55,800 tonnes per annum). (www.internationaloliveoil.org).

This chapter analyze the environmental aspects of table olives sector. After describing the production processes will be analyzed the characteristics of wastewaters, the pollution prevention technologies, and will be evaluate the relative environmental burdens through the LCA methodology.

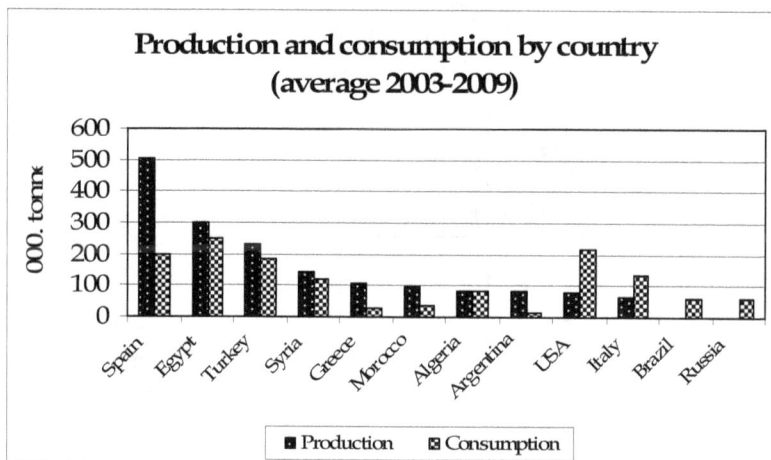

Source: www.internationaloliveoil.org

Fig. 2. Table olives world production and consumption, by distinguishing among producers countries

2. The transformation of table olives

The quality of olives differs from year to year and depends on various things such as climate, rainfall, the amount of pests etc. The methods used for processing olives, according to the IOOC/Codex standard, are listed in Table 1. The degree of ripeness of the drupes when they are picked varies according to the processing method that is going to be used, as does the cultivar – which in some cases has taken on the name of the main area in which the olive-variety is used.

2.1 Green olives
Drupes which are harvested before they are completely ripe, i.e. when they are still green, are intended for processing as "Spanish-style green olives", "Castelvetrano style green olives" or "Naturally-processed black olives".

2.1.1 Spanish-style green olives
The most common method for producing green olives in brine is "Spanish-style" processing. This comprises the following steps: lye treatment (debittering), rinsing, brining, fermentation in brine, packaging and pasteurization.

Table 1. Main methods of processing of table olives

GREEN OLIVES		TURNING COLOUR OLIVES			BLACK OLIVES		
Spanish-style	*Castelvetrano*	*Naturally*	*Naturally (Greek style)*	*Kalamata*	*Backed olives*	*Californian style*
Harvesting, sorting	Harvesting, sorting	Harvesting, sorting	Harvesting, sorting	Harvesting, sorting	Harvesting, sorting	Harvesting, sorting
Size sorting	Size sorting	Size sorting	Size sorting	Size sorting	Size sorting	Size sorting + (preservation in brine)
Lye treatment	Immersion in lye/salt solution	(Incision, cruscing, stoning)		Incision	Dripping	Lye treatment
Washing				Washing		Washing
In brine		In brine	In brine	In brine	Addition of dry salt	Immersion in ferrous salt solution
Fermentation	Fermentation	Fermentation	Fermentation	Immersion in vinegar	Drying oven	(Oxidation to air)
(Size sorting)		Size sorting	Size sorting (+ oxidation to air)			Size sorting
Packaging	Packaging	Packaging	Packaging	Packaging + extravirgin olive oil	Packaging	Packaging
Pasteurization		Pasteurization				Sterilization

Lye treatment (Debittering).

The olives are picked when they have reached their maximum size and are green, or greenish–yellow, in colour; then - after removing any leaves, sorting the drupes and grading them according to size - the olives are treated with an alkaline lye before being immersed in brine. During this treatment the olives are put into a dilute aqueous solution of sodium hydroxide, with a concentration of 1.7% - 4% (w/v); the strength of the concentration depends on the olive-variety, the degree of ripeness of the drupe and the temperature and characteristics of the water used. During the preparation of the lye, the reaction that occurs is exothermic, so the olives should not be put into the solution until it has cooled down. The strength of the concentration of sodium hydroxide depends on the characteristics of the olives being processed: for example, less ripe olives with harder flesh require a higher concentration of soda.

More concentrated solutions can soften the flesh of the drupe, while more dilute solutions - which slow down the debittering process - adversely affect the subsequent fermentation. During treatment the olives must be kept submerged in the solution to prevent oxidation by exposure to the air (resulting in blackening) and to avoid incomplete debittering. In addition, in order to stop the soda from collecting at the bottom of the container (which would result in a solution with varying degrees of alkalinity and thus a non-uniform softening of the drupes), the solution should be mixed and homogenized from time to time (Brighigna, 1998). This debittering phase of the process removes the oleuropein, one of the bitter glucosides naturally present in olive-flesh (Marsilio et al., 2001). The duration of the debittering process ranges from a minimum of 8 hours to a maximum of 15 hours; the treatment is considered complete when the NaOH solution has penetrated from 2/3 to 3/4 of the way into the olive-flesh. Traditionally olives have been treated using fresh lye, but it is possible to re-utilize exhausted lye and thus lower pollution levels (Garrido-Fernandez, 1997; Segovia-Bravo et al., 2008).

Rinsing.

The reason for rinsing the olives with water is to remove most of the lye from the flesh. This phase is very significant from an environmental point of view because it entails the use of large quantities of fresh water, with the consequent production of the same amounts of waste water which contains polluting compounds (Brenes, 2005).

The rinses may be:

- of long-duration: for olives which are meant to be consumed within a short period of time, or for those which are processed as semi-fermented olives; this method involves changing the rinsing water every 8-10 hours for a total duration of 4-5 days in order to obtain an almost clear liquid;

- of medium duration: this is the most commonly-used method. After a first rather short rinse lasting 1-2 hours there are then another 2 rinses lasting 8-12 hours each, resulting in a total of 18-25 hours of rinsing-time. In this case the olives retain enough fermentable substances to ensure proper lactic fermentation;

- of short duration: this consists of a single rinse lasting about 15 hours. With this method the large amount of NaOH solution left in the fruit prevents the rapid onset of lactic fermentation, and sometimes encourages abnormal fermentation. To ensure proper fermentation it is necessary to replace the used brine with fresh, resulting in the consumption of large quantities of fresh water. The elimination of the "brine mother", however, results in the loss of important components such as fermentable substances, minerals, etc.;

-   neutralizing rinses: the aim of this phase is to neutralize the alkalinity of the olives using organic acids (citric acid, ascorbic acid, acetic acid) or inorganic (hydrochloric) acid, or by means of the insufflation of micronized CO_2 into the brine. This procedure reduces water consumption and shortens processing time, while conserving most of the fermentable substances and encouraging a rapid onset of lactic fermentation. Some researchers have looked into the possibility of replacing the traditional rinses with a process which neutralizes the residual alkali using organic or inorganic acids (Brighigna, 1998; Garrido-Fernandez, 1997; Higinio Sánchez Gómez, 2006).

Fermentation in brine.

After rinsing or neutralization, the olives are placed in suitable containers and covered with brine that has a NaCl concentration starting at 9-10% but decreasing rapidly to around 5%, because of the high water content of the olives. Fermentation helps to preserve the product and improve its taste.

It is also possible to add used brines which have been previously analyzed; these are the so-called "brine mothers" which ensure the onset of safe lactic fermentation.

The expression 'onset of fermentation' means the moment when the brine has arrived spontaneously at a pH value of around neutral (± 7), while the fermentation process is deemed complete when, after 2-3 months, the olives are pale yellow in colour and have good texture and a pleasantly-sour taste. If more than 6 months lapses between the stage of fermentation and that of packaging, it is necessary to add salt to the solution in order to stabilize the brine at a concentration ranging from 8% to 10% (Arroyo-López, 2008; Brenes, 2004; Gomez et al., 2006; Garrido-Fernandez, 1997; Hernandez, 2007; Higinio Sánchez Gómez, 2006; Quintana, 1997; Romeo, 2009).

Packaging and pasteurization.

At the end of the production process, the olives are packaged and then pasteurized to prevent progressive deterioration. This operation involves a series of procedures. After the olives are rinsed with fresh water, they are lightly blanched using steam. Then, after sorting to remove any damaged or otherwise defective olives, the containers are filled. The concentration of the brine used for packaging can vary from 3% to 5%, and the pH value must be less than 4.6. After the containers are sealed, they are pasteurized at 90° C for about an hour. (Brighigna, 1998; Javier Casado, 2007; Javier Casado, 2010; Unal. & Nergiz, 2003).

2.1.2 The "Castelvetrano" method

The "Castelvetrano" method, using the olive-variety known as *"Nocellara del Belice"* (typically grown in the Castelvetrano area of Sicily, in Italy, after which the method is named), is used for preparing olives in soda. During this method of preparation, olives which have already been cleaned and size-graded are placed in a solution of water, soda (1.8 -2.5%) and salt (3-5 %). This procedure enables a rapid sweetening of the drupes, which retain a good consistency, a green colour and a distinctive taste of lye due to the NaOH solution (between 0.3% and 0.5%) which is left in the flesh. After 10 to 15 days of debittering treatment, the olives are subjected to a brief rinse using water or brine. The shelf-life of these olives is linked to their pH value, which after 2-3 months tends to go down and thus encourage the development of harmful microorganisms. After packaging, the product should be sterilized in an autoclave at temperatures of about 120° C, to prevent the formation of Clostridium Botulinum and the subsequent production of the Botulin toxin (Brighigna, 1998).

2.1.3 The method used for producing "Naturally-processed green olives"

"Naturally-processed green olives" can be prepared in various ways: whole, crushed, stoned, dressed or flavoured with spices. However, in all these cases the debittering process is carried out naturally, without any form of chemical intervention during the deamarization. In general, it takes at least 10-12 months of fermentation and storage in order to end up with a high-quality product, but there are some varieties of olive which are naturally sweet and require less time (Amelio & De Muro, 2000). The product is preserved in brine, the concentration of which is kept stable between 8% and 10%. The length of the sweetening process depends on several factors such as olive-variety, the amount of oleuropein (a bitter glycoside) in the fruits, the ripeness of the olives and the area where they were grown. The finished product is pleasantly bitter, with a slight winey taste due to the fermentation of the sugar components in the flesh of olives.

The only wastewaters produced during this "natural" fermentation are the rinsing water and the brine. In order to reduce pollution, the brine can be regenerated and re-utilized (Garrido-Fernandez, 1997; Quintana, 1997).

2.2 Black olives

Black olives, which are picked when they are almost fully ripe, may be processed by various methods. In general, the olive-varieties that are grown to be processed as black olives are those whose drupes have a thin peel, flesh with a good consistency and a very good colour, as well as a good flesh-to-pit ratio. When the olives are "naturally processed" in brine, the technology is not very different from that used for green olives (as described above), and the only real difference lies in the organoleptic characteristics of the finished product. However, while the method used for processing "Californian-style" black olives is similar to that used for "Spanish-style" green olives, the amount of alkali, the number of rinses and the colour-fixing stage are all different.

2.2.1 The "Californian-style" method (ripe table olives)

The product of this processing method, originally from California, is thus defined by the US Department of Agriculture: "Olives treated and oxidized during processing, in such a way that they assume a characteristic colour that ranges from dark brown to black, are called 'ripe table olives'". These olives, picked when partially or fully ripe, are first sorted and size-graded.

Storage in brine.

Before treatment with a NaOH solution, the olives are stored in brine at a concentration of between 8% and 10% for at least 30 days. The olives are preserved either in an acid solution (0.4% lactic acid) or by refrigeration (Gomez et al., 2006).

Lye treatment (Debittering)

The original processing method requires repeated debittering treatments (usually three) with a solution of 1%-2% sodium hydroxide, with each treatment lasting between 2 and 6 hours. During the rinsing phases, between one lye treatment and the next, insufflations of air into the water enable the olives to be thoroughly mixed. This helps to darken the surface of the fruit and encourages the enzymatic oxidation of the phenolic compounds present in it. These days the larger olive-processing firms, in order to simplify production, prefer a single treatment using a sodium solution at a concentration of between 1.2% and 1.5%, until the soda has completely penetrated the flesh of the drupes. Agitators or pumps are employed to mix the solution and prevent the soda collecting at the bottom of the container (Brenes, 2004; Higinio Sánchez Gómez, 2006).

The "Californian-style" method has a variant which uses only one debittering treatment with a lye solution at a concentration that ranges from 1.3% to 2.5%. The alkaline treatment is stopped when the lye has penetrated about 2/3 of the way into the flesh of the olives.

Rinsing between treatments

The rinsing, which has to be carried out between the various lye treatments, leads to a significant consumption of fresh water. As mentioned above, the rinses have the dual function of removing the sodium left in the olives and oxidizing them. Olives processed using the variant of this method (with its single debittering treatment) should undergo more rinses. It is also advisable to use lactic acid to sweeten the product and improve its chemical stability.

Immersion in a ferrous salt solution

After the lye treatment, the drupes are immersed for 12-24 hours either in a ferrous gluconate solution (1 - 2 g/L) or in a ferrous lactate solution (0.5-1 g/L). The aim of this phase is to give the olives a uniformly black colour which will be permanent (Brighigna, 1998; Garrido-Fernandez, 1997; Higinio Sánchez Gómez, 2006).

Final rinsing

After immersion in a ferrous salt solution, the olives are rinsed several times (from a minimum of 2 to a maximum of 8) until the rinsing water is pH neutral. Sometimes the water is heated to 80° C to prevent softening of the flesh and its consequent "fish eye" deterioration (caused by gas pockets). Once the rinsing water is pH neutral (after at least 2 rinses) there may be an optional phase of exposing the olives to air for 2-3 days, followed by a further size-grading.

Packaging and sterilization

These olives are usually packaged in brine with 2-5% NaCl and a pH of <4.6. The packaging procedures are basically the same as those for "Spanish-style green olives", described above. The only significant difference is the composition of the brine: in the "Californian-style" method this has a lower salt concentration and a higher pH value, which together create favourable conditions for the development of pathogenic germs and airborne or sporulate bacteria (such as Clostridium botulinum) - with serious consequences both for the consumer and for the product itself. It is therefore necessary to sterilize the containers by subjecting them to temperatures of 121° C for about an hour. After this heat treatment, samples from each batch should be subjected to microbiological controls in a specialized laboratory. It must be said, however, that sterilization affects the organoleptic properties of olives. Moreover, in some olives treated in this way, the flesh becomes less firm. (Kanavouras, 2005). The injection of calcium chloride (1 kg per 100 kg of olives) into the final wash may help to maintain the consistency of the drupes; any use of this additive, however, must be explicitly mentioned as 'E 509' on the label of the container.

3. Chemical characteristics of the wastewater from table olives processing

The production of table olives (green, naturally black etc.) involves various and consistent flows of wastewater (from min 0.5 liters/kg to 6 liters/kg). Every year, in the world from almost 1 million to almost 11.7 million tons of wastewater are generated from the processing of 1.9 million tons of table olives (assessment done by the author on the 2003-2009 average data coming from IOC, International Olive Council). Table 2 shows the production and assessment of the wastewater generated from the table olive processing, distinguishing for Country. The number of flows and their respective volumes are different and they depend of the kind of finished product. In order to have a picture more complete, the wastewater deriving from the washes of the container and those used for the packaging should be added to the amount indicate in the table 3. The processing of table olives dates back to a lot of years ago, so the industries never considered the availability of the water as resource and the environmental effects deriving from its use. The processing of table olives is an activity concentrated in a few months per year (in particular autumn-winter) and in restricted geographic areas (sometimes with little surface water resources). This determines strong pressure on the water resources and on the quality of the surface water.

Before that these problematic were pointed out with particular gravity, relevance wasn't been given to the surface sampling, or at the ground and at the draining modality of the wastewater. The growing attention towards the use of natural resources and the arising of the national laws and, in the case of Europe Union, of community laws ever more constricting for what concerning the draining of wastewater, involve the needing to assess with attention the modality of the water use into the table olives production processes.

	Countries										
	ES	EG	TR	SY	GR	MA	DZ	AR	USA	IT	Others
Production	503.3	299.6	230.8	143.3	108.2	98.3	81.9	80.8	79.7	65	249
Estimated wastewater (min)	252	150	115	72	54	49	41	40	40	32	125
Estimated wastewater (max)	3020	1798	1385	860	649	590	491	485	478	390	1494

Source: Author estimation based on data COI

Table 2. Table olives production and estimated waste water (average 2003-2009) (1,000 tonnes)

Table olive processing methods	Lye	Fermentation brine	Washing	Preservation brine	Total
Spanish style	0.5	0.5	0.5-2.0	0.5	2-3.5
Untreated green and turning colour olives		0.5		0.5	1
California green ripe olives	0.5	0.5	0.5-2.0	0.0-0.5	1.5-3.5
California black ripe olives	0.5-2.5	0.5	0.5-3.0	0.5	2-6.5
Naturally black olives		0.5		0.5	1

Source: Garrido-Fernandez et al, 1997

Table 3. Volume (L) of wastewater per kg of olives produced during the main phases of different types of table olives.

The problem can be cope with different and complementary ways:
- By trying to decrease the entire water requirement in the process (by reduction of the volume used in the single step of the process, exclusion of some operations from the process or reduction of its number, reuse of the same water on the same operation but on different batches of table olives;
- By trying to reduce the environmental impact of the wastewater (low percentage of NaOH in the lye or NaCl in the fermentation brine);
- By setting specific wastewater treatment methods.

The approaches listed above in point 1 and 2 are process improvements, while, the wastewater treatment is not affected by modifies applied to the processing method.

All these approaches were been studied in the last ten years. However, before to analyzing the most important results, it could be useful to watching the chemical characteristics of the wastewater generated by the various processes.

3.1 Wastewater from Spanish-style green olives processing

The approximate characteristics of the wastewaters from this treatment of the green olives are reported in the table 4.

General characteristics of wastewaters from processing Spanish-style pickled green olives in brine			
Characteristic	Lye	Washing water	Fermentation brine
pH	9.5 – 12.0	9.0 – 11.5	3.8 – 4.2
Free NaOH (g/L)	11.0	1.5	-
Free acidity (g lactic acid/L)	-	-	6 – 15
Polyphenols (g tannic acid/L)	2.5 – 4.0	2.5 – 4.0	4.0 – 6.0
Reducing sugars (g glucole /L)	6 -9	6 -9	-
Dissolved organic solids (g/L)	20 - 30	20 - 30	15 - 25
Dissolved inorganic solids (g/L)	20 - 35	7 - 25	90 - 110
COD (g O_2/L)	15 - 35	12 - 35	10 - 35
BOD$_5$ (g O_2/L)	9 - 20	9 - 20	8 - 20

Source: Garrido-Fernandez et al, 1997

Table 4. Characteristics of the wastewater from Spanish-style green olives processing

All these wastewaters are highly polluting and are not simply treated by conventional methods. As it can be seen, there are two different groups of wastewaters, alkaline (lye and washing waters) and acidic (fermentation brine). Their management and treatment need to be separate and different.

Lye and washing water noticeable differ in residual alkali concentration, and, in minor amount, in dissolved inorganic solids. The content of polyphenols is higher in the fermentation brine, while is null the one of reducing sugars.

Longer is the contact time between fruits and solution in each step greater are the dissolved substances. So, if it is good to dissolve the greatest amount of polyphenols, it is not good to extract reducing sugars during the lye treatment and washing. They are necessary during the successive fermentation step.

The brine' main characteristic is the high concentration of inorganic components (mainly Na and K salts). They come from the added NaCl and from the high content of the olive flesh.

The contemporary presence of substances easily used by microbes as carbon sources (reducing sugars in lye and washing waters, lactic acid in brine where it accounts for about

80% of the BOD$_5$), and substances that interfere with them (polyphenols), seriously interfere with the direct application of any biological purification procedure.

In conclusion, these wastewaters are heavily polluted, perhaps difficult to treat and dispose of. Many strategies have been studied to reduce the environmental impact of these wastewaters: internal control measures, such as lye and washing waters re-use, reduction or elimination of washing waters, debittering with low-concentration lyes, regeneration and re-use of fermentation brine. Any of these approaches has completely resulted in meeting the needs (Garrido-Fernandez et al., 1997).

Table 5 shows the value, expressed in grams per litre of oxygen, biological oxygen demand (BOD$_5$) and chemical oxygen demand (COD) of wastewater processing from Spanish-style.

Wastewater	pH	BOD$_5$ (g O$_2$/L)	COD (g O$_2$/L)	NH$_3$ (mg/L)	B (mg/L)	P (mg/L)	Fe (mg/L)	Cu (mg/L)	Zn (mg/L)	Ni (mg/L)	Hg (mg/L)	Na (g/L)	K (g/L)	Cl- (mg/L)
Lye	12.7	4.8	24.0	1.3	0.14	18.3	5.75	0.23	0.48	0.12	0.08	96.5	18.5	2.9
1st Washing	9.1	3.1	10.0	16.8	0.63	28.4	3.62	0.16	0.36	0.09	0.05	15.7	4.2	1.9
2nd Washing	7.6	1.6	7.6	1.3	0.40	16.3	1.54	0.10	1.64	0.03	0.05	10.5	2.6	1.1
3rd Washing	6.2	0.9	3.0	2.0	0.34	11.4	0.84	0.06	0.08	0.02	0.09	5.0	1.7	0.7
4th Washing	6.5	0.6	2.0	2.3	0.28	10.6	0.82	0.04	0.07	0.01	0.06	3.7	1.1	0.5
5th Washing	7.0	<0.1	0.2	3.5	0.16	4.8	0.40	<0.01	0.12	0.01	0.05	1.1	0.3	0.2

Source: Garrido-Fernandez et al, 1997

Table 5. Loads polluting wastewater processing of green olives by the "Spanish-style"

3.2 Wastewater from turning colour olives processing

Processes not based on an alkaline treatment, comprising only fermentation or storage brine and, sometimes, water from washing before packing, produce a less volume of wastewaters that are less contaminated.

Characteristic	Main wastewaters from a typical darkening stage							
	Holding brine	1st treatment		2nd treatment		3rd treatment		Gluconate solution
		Lye	Wash	Lye	Wash	Lye	Wash	
pH	-	12.11	6.90	11.85	8.00	11.70	8.55	3.70
Free NaOH (g/L)	-	8.90	-	4.60	-	4.30	-	-
Dissolved inorganic solids (g/L)	-	2.38	3.18	4.34	7.21	6.88	8.14	7.33
Dissolved organic solids (g/L)	-	19.27	3.15	12.25	32.21	12.08	36.64	43.38
Polyphenols (g/L)	-	0.18	0.22	0.59	0.31	0.48	0.64	0.80
COD (g O$_2$/L)	35.0	2.0	3.5	2.5	3.8	2.6	3.7	1.5

Table 6. Characteristics of the wastewater from turning color olives processing

Process that includes alkali treatment and the darkening stage (by the use of ferrous gluconate) (California-style method), needs a greater volume of water, due mainly to the numerous lye treatments and washes (Table 6).

It can be seen that the highest pollutant charge comes from the holding brine, while lyes and washing waters are relatively moderate. Fermentation-storage brine has characteristics very similar to those from Spanish-style green olives (Garrido-Fernandez et al, 1997).

3.3 Wastewater from naturally black olives processing

Fermentation brines are the only waste liquids produced during this type of processing that pose some problem. They are liquids with a high level of organic matter (Table 7).

Characteristics	Brines from different origins
pH	3.6 - 4.4
Salts (g NaCl/L)	56.0 - 77.0
Dissolved organic solids (g/L)	95.3 - 118.8
BOD_5 (g O_2/L)	34.8 - 38.3
Polyphenols (g/L)	3.2 - 5.1

Table 7. Characteristics of the wastewater from naturally black olives processing

It is the highest between all types of table olives processing, very close to those reached in olive oil mill wastewaters. The molecules contained in the organic matter of these brines, however, are only heavy potential contaminants, as shown by their high BOD_5 .

The salt content is higher than in any other process. Such concentrations can cause serious problems for conventional municipal wastewater treatment facilities.

3.4 Wastewater from black olives by Californian - style

Tables 8 and 9 show the values of BOD_5 and COD of wastewater processing of black olives by Californian-style and its variant, table 10 the values of TOC of packing brines of Spanish-style and Californian-style, and Table 11 the characteristic of Californian style wastewater.

Wastewaters	BOD_5 (mgO$_2$/L)	COD (mgO$_2$/L)
Preservation	10,620	35,000
1st Lye treatment	607	2,000
1 st Washing	1,062	3,500
2nd Lye treatment	759	2,500
2nd Washing	1,153	3,800
3rd Lye treatment	789	2,600
3rd Washing	1,123	3,700
Immersion in ferrous salt solution	455	1,500
4 th Washing	394	1,995
5th Washing	394	1,995

Source: Garrido-Fernandez et al, 1997

Table 8. Pollution loads of wastewater from processing of black olives processed by the Californian style

Wastewaters	BOD$_5$ (mgO$_2$/L)	COD (mgO$_2$/L)
1 st Lye treatment	150	4,260
1 st Washing	230	6,210
2 nd Washing	200	3,880
3 rd Washing	190	1,480
4 th Washing	190	1,680
5 th Washing	294	2,950
6 th Washing + lactic acid	336	5,550
7 th Washing + lactic acid	200	5,840
8 th Washing + lactic acid	434	5,580
Immersion in ferrous salt solution	313	6,650
9 th Washing	430	2,320
10 th Washing	357	1,670

Table 9. Pollution loads of wastewater from black olive processing by variant of the Californian style

Preservation brine	Green olives	Black olives
TOC (mg/L)	3,089	3,368

Table 10. Loads of pollutants brines packaging, expressed as total organic carbon (TOC)

"Californian Style" wastewater	
Method A	COD
1 st Lye treatment	2,000
1 st Washing	3,500
2 nd Lye treatment	2,500
2 nd Washing	3,800
3 rd Lye treatment	2,600
3 rd Washing	3,700
Immersion in ferrous salt solution	1,500
4 th Washing	1,995
5 th Washing	1,995

Source: Garrido-Fernandez et al, 1997

Table 11. Characteristic of Californian style wastewater

4. Pollution prevention methods in the table olive industries

Prevention is better than curing. So internal control measures of potential pollutants or water use must be preferred as the first choice to tackle the environmental impacts of these industries.

Internal control measures include:
- Re-use of lye, washing waters, fermentation/storage brine;
- Reduction or elimination of washing waters;
- Debittering with low-concentration lyes;
- Salt-free storage solution.

Notwithstanding many experimental studies fulfilled during the latter decades, no one has produced results totally sharable, so, not even one was introduced as process improvement. This because, each modify, introduced respect to the traditional process, ever involves variations in the quality of the product which aren't accepted. As was noted, the chemical composition of the table olive wastewaters limits the effectiveness of classical purification systems. Polyphenols (oleuropein and derivates, anthocyanins) and sodium content (from the NaOH in the lye and from NaCl in the brines) are the most important pollutants contained.

Accordingly, in the latter years a number of studies were carried out in order to adapt these wastewaters to the limits accepted by the national environmental laws (Deligiorgis et al., 2008).

Purification (partial) of lyes and washing waters from green table olives processing by lowering the pH under the neutrality and by leading the temperature of the wastewater around $0°$ C (De Castro et al., 1983), allow to bring COD down over 35%. The wet oxidation carried out in alkaline environment into a pressurized reactor, at the temperature of $200°$ C, is able to eliminate almost all over the polyphenols, and if the pure oxygen is used, also the 90% of the organic carbon present at beginning.

The efficiency is lower when the same methods are applied to the washing waters from the darkening process of black-ripe olives. Fermentation brine represents more than the 22% of the total final wastewaters from green table olives processing, but account for approximately 70% of the organic pollution charge. Thus, if these brines are correctly purified much of the pollution potential is eliminated. Physicochemical treatment reduce the pollutant charge (COD) of the fermentation brines only about 20÷25%.

There is not much experience of biological treatment to remove the excess organic charge up to now (Beltran et al., 2008). The main problem is related to polyphenols degradation and the salt level; appropriate dilution and/or adapted microorganisms are needed. *Brenes et al.* showed that an activated sludge process can be used successfully, yealding a 80% COD reduction, but only a small proportion of polyphenols was consumed. A manner to reduce the polyphenols content could be the wet air oxidation (Brenes et al., 2000; Beltràn et al., 2000a; Katsoni et al., 2008).

The best results were obtained at acidic pH, catalized by means of Cu_2+. The addition of H_2O_2 in the absence of copper resulted in a lower COD conversion, while an increase in the biodegradability of the final mixture was found after having added the radical promoter.

Kotsou et al. studied an aerobic biological treatment using an Aspergillus niger strain, in combination with chemical oxidation (H_2O_2), followed by coagulation with CaO of the resulting treated waters. The results of the experiments are shown in the Table 12 (Kotsou et al., 2004).

The main effect of the chemical oxidation step was the elimination of persistent phenolic compounds during the biological treatment of total phenolic compounds.

Influent		After the biological step HT 2 days		After chemical Oxidation [H_2O_2] 4g/L [Fe^{2+}] 0.5 g/L		Affluent after liming CaO 5 g/L Sed. Time 2h	
COD mg/L	Phenols mg/L	COD mg/L	Phenols mg/L	COD mg/L	Phenols mg/L	COD mg/L	Phenols mg/L
11000	190.7	3950	62.4	1537	32.4	835	5.4
11130	185.5	4043	64.7	1625	29.5	825	5.1
9300	150.8	3575	38.1	1327	25.7	790	4.8
8500	147.9	3616	39.6	1452	17.4	890	4.3
20920	319.5	5185	187.5	2225	63.2	1675	21.2
9750	157.8	3275	30.7	1297	27.3	675	4.3
9080	152.2	2135	33.2	1050	16.9	780	4.6
8190	137.4	2700	32.3	1027	11.8	795	3.4
9960	178.3	2160	41.6	1097	20.1	835	3.8
9850	185.3	2227	37.4	1105	22.7	895	3.5

Source: Kotsou et al., 2004

Table 12. Characteristics of the wastewater after aerobic biological treatment

Beltran et al. studied the purification efficacy of the aerobic and anaerobic biodegradation of the wastewater from green table olive processing using acclimatized bacterial flora taken from, respectively, an activated sludge from a municipal wastewater treatment plant and a biomass from an anaerobic digester of a municipal wastewater treatment plant.

The aerobic biodegradation a significant reduction of the COD between 50% and 70%, and an important removal of the total polyphenolic compounds around 97%; the anaerobic digestion achieves an important removal of COD between 81 and 94%, indicating that most of the substrate fed to the digester is biodegraded anaerobically (Beltran et al., 2008).

Different chemical oxidants, alone and combined, were applied to the purification of the wastewaters from the storage in brine of black table olives. The most effective purification process was the overall combination O_3/UV/ H_2O_2. Aerobic treatment of the effluents gave a major substrate removal that was independent of the initial biomass concentration. The chemical treatments in general, and ozonation processes in particular, are useful for the degradation of organic matter, especially aromatic compounds.

These processes can be used as pre-treatment steps for subsequent aerobic degradation in order to meet discharge norms and reach purification efficiencies required by national regulations (Beltran-Heredia et al., 2000a; Beltran-Heredia et al., 2000b; Beltran-Heredia et al., 2000c; Beltrán de Heredia et al., 2001d; Beltrán de Heredia et al., 2001a; Beltrán de Heredia et al., 2001b; Benitez et al., 2003).

The figure 3 shows the results obtained in experiments carried out on the wastewater of black ripe table olives which foresaw only the aerobic biologic treatment (1), the only treatment by ozone (2), the treatment by ozone followed by a aerobic biologic treatment (3), the aerobic biologic treatment followed by the ozone treatment (4).

Figure 3 shows that the biologic treatment is more efficient in order to reduce the COD, while the ozone is more efficient in the reduction of the polyphenols. The two treatments employed one after the other allow to bring down almost 90% of COD and polyphenols. The wastewater treated are odorless, uncolored and disinfected. They could be reintroduced into another productive loop.

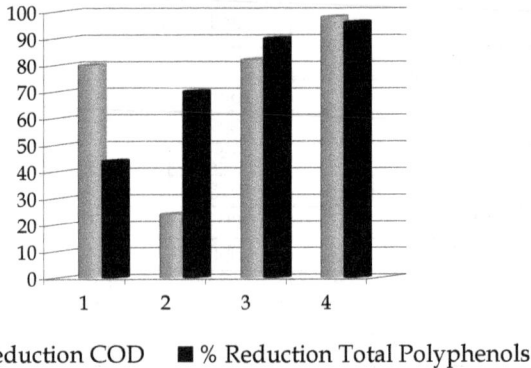

% Reduction COD   ■ % Reduction Total Polyphenols

Source: Beltran-Heredia, 2001

Fig. 3. Results obtained by ozone treatment on the wastewater of black ripe table olives

Instead of to eliminate the polyphenols from the washing water, an alternative washing process could be that which foresee the recovery of the substances useful from an economic point of view (e. g. hydroxytyrosol) (De Castro & Brenes, 2001; Bouaziz et al., 2008).

A similar proposal, but referred to the exhausted lyes deriving from the deamarization phase of the various table olive processing methods, is in the research project of D'Annibale et al. "New technology in virgin olive oil mechanical extraction process in relationship with the possible traceability and nutritional and sensory quality of oil and exploitation of vegetation waters and pomaces by recovery the bioactive products" (D'Annibale et al., 2005). This proposal is enriched from the proposition to produce enzymes from brines in order to create an economic interest. When, for various reasons, it is impossible to apply some wastewater depuration treatment able to bring down the pollution in order to dispose of it in agreement with laws, in the warm regions in which the weather is windy and there is drought, an extreme remedy could be the lagoon. In this case great attention should be given to the bad smells and leaching risk. The evaporation can be increased, favouring the absorption of the sun radiation into the lagoon basin (Chatzisymeon et al., 2008), by using surfactant agents, re-circle and spraying.

5. Life Cycle Assessment (LCA) methodology to assess the impact of wastewater of table olives industries

5.1 About Life Cycle Assessment

The LCA (Life Cycle Assessment), is a valuable analysis tool, potentially applicable to any product, process or service, and it is subject to specific standards by the International Standardization Organization (ISO). The rules of the ISO 14040 series of the Environmental Management Life Cycle Assessment (four standards dedicated each one to a specific part of the methodology), it is the point of reference for the application, in business decisions, of this environmental management methodology (ISO 14040, 2006; ISO 14044; ISO /TR 14048). The LCA can be used as a technical tool to identify and assess opportunities to reduce the environmental effects associated with specific products, production processes, packaging, materials or activities or to identify various scenarios on which to make strategic choices. This tool can also be used to achieve a sustainable management of the natural resources. This methodology, adopting a systemic approach, can be used as tool of private or public decisions in the various productive choices during a definition of environmental standards, to establish the basis of information from a single stage of the production process up to chain, to identify "process hot spots", as a product certification tool (for example Eco-label). The LCA, considers the product as a system, that is analyzing the changes and flows of matter and energy, since their withdrawal from national system, that of production of the product, until the final disposal.

Numerous LCA studies were focused on the olive oil chain (Notarnicola et al., 2004; Olivieri et al., 2007; Raggi et al., 2000; Olivieri et al., 2005; Russo et al., 2008), but not many on the table olive chain (Cappelletti et al., 2010), however each business reality presents different characteristics (Salomone et al., 2010).

5.2 The goal and scope definition

The Life Cycle Assessment methodology was applied to the various methods used to process green and black table olives.

As for the green table olives the processes analysed were: Spanish- style, Castelvetrano-style and Green natural-style, while, as far as the black table olives, two methods of the California black ripe olives were studied (Russo et al., 2010). This analysis aims to highlight the contribution, of the wastewater produced from each processing method, to the totally environmental burden referred to the table olives industries. For this reason the system boundaries were simplified including only the sub-phases referred to the processing methods. In order to better focusing the study on the industrial phase, the agricultural phase and packaging were excluded from the analysis. The Functional Unit (FU) chosen was 100 kg of fresh table olives.

5.3 The Life Cycle Inventory (LCI)

As for the input and output concerning the processing methods, all the relevant data were considered. Although in different quantity, Spanish-style (figure 4) and Castelvetrano-style (figure 5) have the same input, while the input of green natural-style (figure 8) are quite different for quantity and type respect to these of the other methods used to processing green table olives. As for the methods used to processing black olives, the two methods of California black ripe olives present same input (figure 6 and 7), but also in this case the amount is not the same. As for the output, the amount of waste are the same for the green olives processes and quite high for the two methods of California black ripe olives. As far as the wastewater, figure 9 shows the quantity produced by each processing methods; among the various processes, the method B of the California black ripe olives produces the greater amount of wastewater, almost 9 kg per kg of fresh raw material. This is due to the lower quantity of caustic soda used respect the method A and the higher number of washes needed to oxidizing the olives and fixing their black colour.

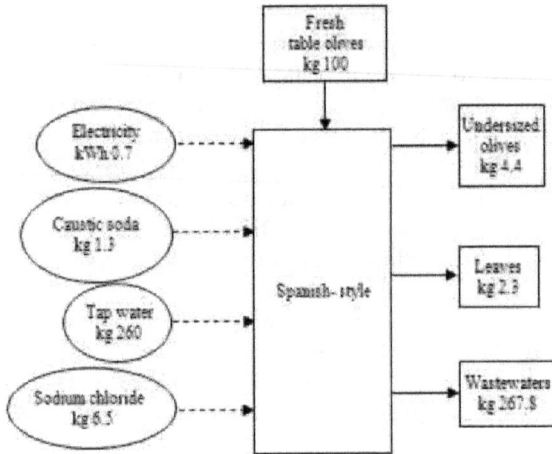

Fig. 4. Layout of Spanish-style

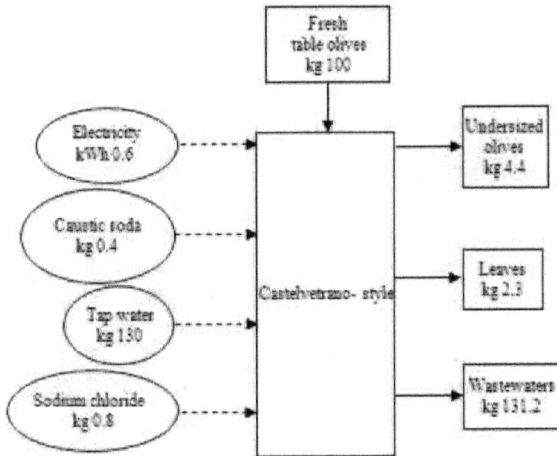

Fig. 5. Layout of Castelvetrano-style

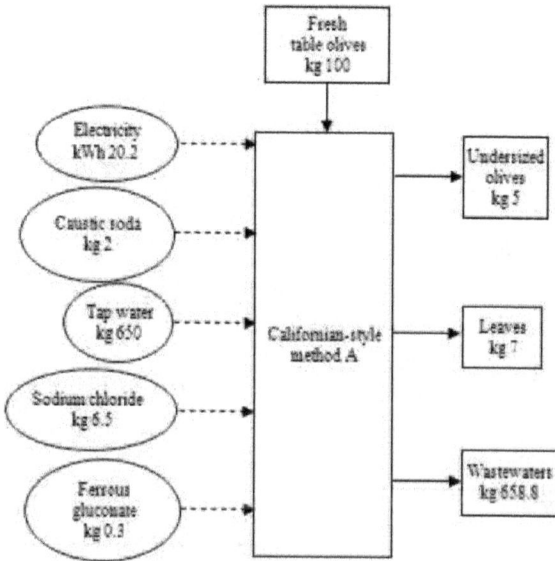

Fig. 6. Layout of California black ripe olives method A

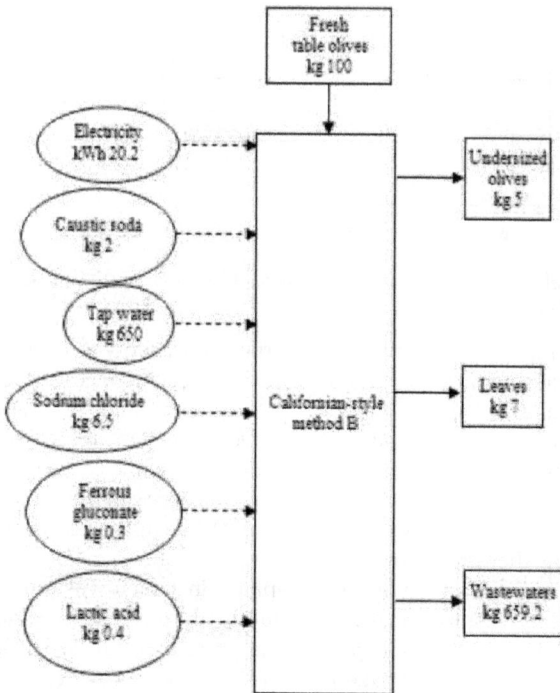

Fig. 7. Layout of California black ripe olives method B

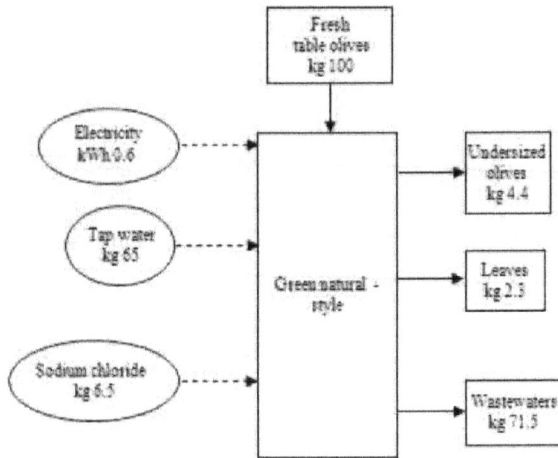

Fig. 8. Layout of the green natural-style

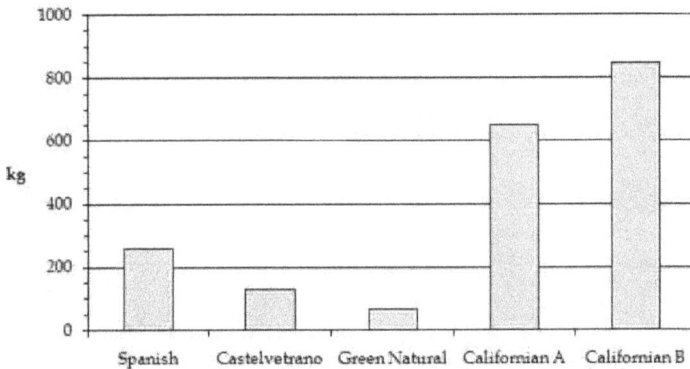

Fig. 9. Wastewater referred to 100 kg of processed table olives, distinguishing among the various manufacturing processes

In order to better compare the amount of materials and energy and the waste produced by the processing methods considered, table 13 shows, for each one, the amount of all the input and output. In particular, as for the wastewater, these are distinguished for each sub-phases. Among the methods used to processing green olives, the Spanish-style is the one that needs the higher amount of water, electricity and caustic soda, while comparing all the processes, the method B of the California black ripe olives needs more quantity for almost all the input (exception for caustic soda, employed in higher quantity in the California black ripe olives method A). Focusing the attention on the production of wastewater, table 1 shows that, exception for the green natural style, the sub-phase in which wastewater are produced in greater quantity is the washing phase. It is due to the need to eliminate the residues of caustic soda after the deamarization phase. The green natural-style don't employ caustic soda because of the deamarization take place in a solution of water and salt and after the olives don't need to be washed, but this method take a long time and not for all the variety of table olives is convenient from an economic point of view.

		Green table olives			Black table olives	
Input		Spanish style	Castelvetrano style	Green Natural -style	California black ripe olives method A	California black ripe olives method B
Water	kg	260.0	130.0	65.0	650.0	845.0
Electricity	kWh	0.7	0.6	0.6	20.2	27.3
Sodium chloride NaCl	kg	6.5	0.8	6.5	6.5	6.5
Lactic Acid	kg					0.4
Caustic Soda NaOH	kg	1.3	0.4		2.0	1.3
Ferrous Gluconate	kg				0.3	0.3
Output						
Undersized Olives and Leaves	kg	6.7	6.7	6.7	12.0	12.0
WW Preservation in Brine	kg				71.5	71.5
WW Deamarization	kg	66.3	66.2	71.5	197.0	66.3
WW Washes	kg	130.0	65.0		325.0	650.0
WW Fermentation	kg	71.5				
WW Ferrous gluconate attack	kg				65.3	65.3

Table 13. Input and output referred to the various table olives manufacturing processes

5.4 Data quality

The inventory data was collected from various sources. The data concerning the input of technological processing was collected directly from processing companies which produce green table olives, by using Spanish-style, and black-ripe table olives, by using California black ripe olives method A and B. Electricity consumption was measured, as well as the amount of input resources and wastewater. For what concerning the green natural-style and Castelvetrano-style, also some data coming from literature was considered (Garrido-Fernandez et al., 1997). As far as electricity production, the Italian mix was the type considered (AEEG, 2010). As far as the issues of wastewater; by means of laboratory analyses we assessed the pollution caused by the exhausted brines, lyes and washing waters referred to the Spanish-style and California black ripe olives, while for green natural-style and Castelvetrano-style we considered literature data. All collected data was processed by using GaBi4 software and its databases (IKP and PE, 2002), also Ecoinvent database was used (Frischknecht et al., 2004).

5.5 LCIA (Life Cycle Impact Assessment)

The CML 2001 impact assessment method was used to analyse the environmental impact of the input and output measured during the inventory phase. The categories of pollution considered by this method were: abiotic depletion potential (ADP), acidification potential

(AP), eutrophication potential (EP), freshwater aquatic ecotoxicity potential (FAETP inf.), global warming potential (GWP 100 years), human toxicity potential (HTP), marine aquatic ecotoxicity potential (MAETP), ozone layer depletion potential (ODP, steady state), photochemical ozone creation potential (POCP), radioactive, radiation (RAD) and terrestric ecotoxicity potential (TETP). All these impact categories were opportunely weighted and normalized in order to obtain an eco-indicator.

Figure 10 shows, for each processing method, the value of the CML 2001 eco-indicator deriving from the sum of the normalized value of the impact categories. As for the methods used to processing green olives, the category EP is the most important, while as far as the California black ripe olives the category that mainly contribute to the total eco-indicator is MAETP, but also EP and GWP increase the value. This is due to the impact deriving from the greater use of the electricity required to oxidize the olives and fixing their black colour.

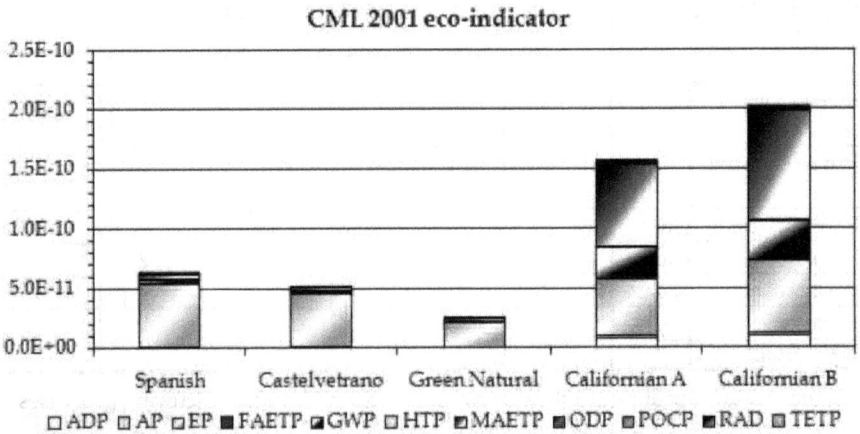

Fig. 10. Eco-indicator referred to the various table olives manufacturing processes

The LCA results underlined that the eutrophication is a very important impact for the table olives processing industries, and it derives from the pollution of the wastewater. In order to better understand the pollution of the wastewater of table olives industries, figure 11, shows, for each processing method the contribution to the eutrophication of the wastewater produced in the different sub-phases of the process.

By analysing the Spanish-style, the exhausted brine is more pollutant than lye, while, even though the washing waters are greater in volume, they are the less pollutant, the deamarization phase of the Castelvetrano-style is quite different from that of Spanish-style because sodium chloride is added to the lye, in order to accelerate the deamarization process, so, as shown in figure 11, the wastewater from this phase have higher environmental burdens, the washing waters further affect the eutrophication, but the total value is lower than the one referred to the Spanish-style.

As for green natural style, the process don't foresee washes but only deamarization in brine, so the wastewater produced in this phase are the only contributor to the EP and their pollution, among all the processes analysed is the lowest.

By analysing the California black ripe olives, figure 11 shows that as for the EP an important contribution is due to the wastewater deriving from the phase of preservation in brine, also the washing waters heavily contribute to the total EP, especially in the method B, while the

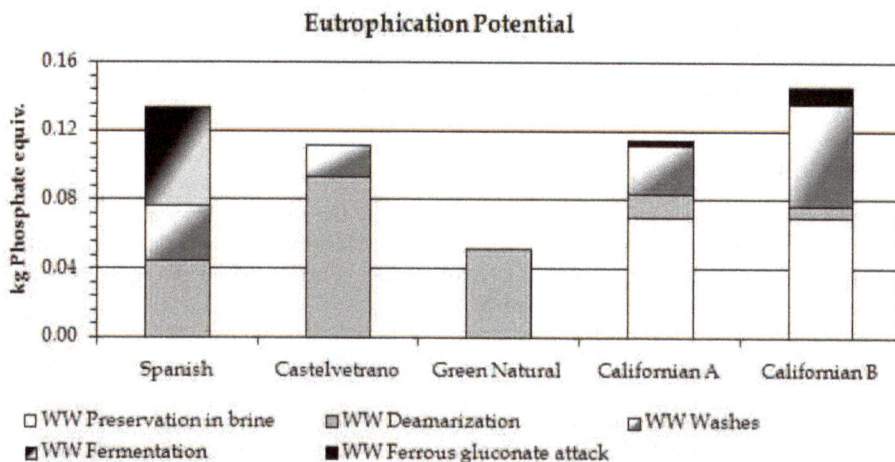

Fig. 11. Eutrophication Potential, distinguishing among the various table olives manufacturing processes

exhausted lyes and the wastewater deriving from the phase of ferrous gluconate attack present a low environmental impact.

Finally, even though the processes compared make use of different technologies to obtain end products each with distinct organoleptic characteristics, we can state that "Green natural style" processing gives a better environmental performance in terms of eutrophication potential and the California black ripe olives method B appears to be the most polluting overall, but it is relevant to point out that method A of California black ripe olives has a lower EP value than Spanish-style.

5.6 Improvement hypothesis

The Life Cycle Assessment has highlighted that the wastewater is an important issues deriving from table olives industries, and the impact category mainly affected by the pollution of wastewater is the eutrophication. Among the various processing methods used to process green and black table olives, wastewater are produced in different amount and affect in different way the impact category EU.

In order to improve the environmental performance of these processing–system, however, technological solutions could be adopted with the aim of reducing the amount of material used (caustic soda, sodium chloride, water), and consequently limit the negative effects connected to its production and use (Marsilio et al., 2008). It is also worth considering reusing the brine and the rinsing-water, in order to reduce the amount of wastewater produced and the emissions (Segovia-Bravo et al., 2008). An important improving solution could be also represented from the extraction of the useful substances, such as phenols, responsible to the high pollution of the wastewater (De Castro et al., 1983; Garrido-Fernandez, 1983; Garrido Fernandez et al., 1997; Bouaziz et al., 2008). Finally, especially for the Californian methods that use greater amount of electricity, environmental improvement should concern also the minimizing of the impacts deriving from energy production. This objective could be achieved by, e.g., installing renewable energy sources at the factory in order to use the electricity directly for these uses. It is important to consider that the adoption of these solution made with regard to achieving the objectives of eco-compatibility, leads to advantages, which are not only environmental (a lower consumption of resources, a reduction in pollutants etc.) but also economic (i.e. lower costs).

Contribution of authors

This paper has been thought, discussed and written by the three authors and it is the result of their common commitment, in particular Giulio Mario Cappelletti has contributed to paragraphs 1 and 2; Giuseppe Martino Nicoletti has contributed to paragraphs 3 and 4; Carlo Russo has contributed to paragraph 5.

6. References

AEEG, www.autorita.energia.it (Website consulted Sep 2010).

Amelio, M. & De Muro, E. (2000). Naturally fermented black olives of Taggiasca variety (Olea europaea L.). *Grasas y Aceites*, 51, 6, 429-439.

Arroyo-López, F.N.; Querol, A. ; Bautista-Gallego, J. & Garrido-Fernández, A. (2008). Role of yeasts in table olive production. *International Journal of Food Microbiology*, 128, 2, 189-196.

Beltran-Heredia, J.; Torregrosa, J.; Dominguez, J. R. & Garcia, J. (2000a). Treatment of black-olive wastewaters by ozonation and aerobic biological degradation. *Water Research*, 34, 14, 3515-3522.

Beltran-Heredia, J.; Torregrosa, J.; Dominguez, J. R. & Garcia, J. (2000b). Ozonation of black-table-olive industrial wastewaters: effect of an aerobic biological pretreatment. *Journal of Chemical Technology & Biotechnology*, 75, 7, 561-568.

Beltran-Heredia, J.; Torregrosa, J.; Dominguez, J. R. & Garcia, J. (2000c). Aerobic biological treatment of black table olive washing wastewaters: effect of an ozonation stage. *Process Biochemistry*, 35, 1183-1190.

Beltràn, F. J. ; Rivas, F. J. ; Gimeno, O. & Frades, J. (2000d). Wet air oxidation of wastewater from table olive elaboration industries, *Proceedings of the Symposia of the Division of Environmental Chemistry of the American Chemical Society*, San Francisco (CA), March 26-30.

Beltrán de Heredia, J. A.; Torregrosa, J. A.; Vargas, J. R. D. & Garcìa Rodrìguez, J. (2001a). Depuración de las aguas de lavado de aceitunas negras mediante procesos biológico aerobio y de ozonización. *Grasas y Aceites*, 52, 3-4, 184-191.

Beltrán de Heredia, J.; Torregrosa, J.; Domínguez, J. R. & García, J. (2001b). Tratamiento con ozono de lejías residuales del aderezo de aceitunas negras. *Grasas y Aceites*, 52, 1, 17-25.

Beltran, J.; Gonzalez, T. & Garcia, J. (2008). Kinetics of the biodegradation of green table olive wastewaters by aerobic and anaerobic treatments. *Journal of Hazardous Materials*, 154, 1-3, 839-845.

Benitez, J. F.; Acero, J. L. & Leal, A. I. (2003). Purification of storage brines from the preservation of table olives. *Journal of Hazardous Materials*, B, 96, 155-169.

Bouaziz, M.; Lassoued, S.; Bouallagui, Z. , Smaoui, S.; Gargoubi, A.; Dhouib, A. & Sayadi, S. (2008). Synthesis and recovery of high bioactive phenolics from table-olive brine process wastewater. *Bioorganic & Medicinal Chemistry*, 16, 9238-9246.

Brenes, M.; Garcìa, P.; Romero Concepciòn & Garrido, A. (2000). Treatment of green table olive wastewaters by an activated-sludge process. *J. Chem. Tech. Biotech.*, 75, 6, 459-463.

Brenes, M. (2004). Olive Fermentation and processing: scientific and technological challenges. *Journal of Food Science*, 69, 1, FMS33-FMS34.

Brenes, M. & García, P. (2005). Elaboración de aceitunas denominadas «Green ripe olives» con variedades españolas. *Grasas y Aceites*, 56, 3, 188-191.

Brighigna, A. (1998). *Le Olive da tavola*, Edizioni agricole, Bologna, Italy, 1998.

Cappelletti, G. M. ; Nicoletti, G. M. & Russo, C. (2010). Life Cycle Assessment (LCA) of Spanish-style green table olives. *Italian Journal of Food Science*, 22, 1, 3-14.

Chatzisymeon, E.; Stypas, E.; Bousios, S.; Xekoukoulotakis, N. P. & Mantzavinos, D (2008). Photocatalytic treatment of black table olive processing wastewater. *Journal of Hazardous Materials*, 154, 1090-1097.

D'Annibale, A. et al. (2005). New technology in virgin olive oil mechanical extraction process in relationship with the possible traceability and nutritional and sensory quality of oil and exploitation of vegetation waters and pomaces by recovery the bioactive products (retrieved from
http://www.ricercaitaliana.it/prin/dettaglio_prin_en-2005070135.htm)

De Castro, G. M. ; Duran Quintana, M. C. ; Garcia Garcia, P., et al. (1983). Processing fruits of «Gordal» variety, without washing and reusing lye solutions, as Spanish style green style olives. Wastewaters study and some purification tests. *Grasas y Aceites*, 3, 162.

De Castro, A. & Brenes, M. (2001). Fermentation of washing waters of Spanish-style green olive processing. *Process Biochemistry*, 36, 797-802.

Deligiorgis, A.; Xekoukoulotakis, N. P.; Diamadopoulos, E. & Mantzavinos, D. (2008). Electrochemical oxidation of table olive processing wastewater over boron-doped diamond electrodes: Treatment optimization by factorial design. *Water Research*, 42, 1229-1237.

Frischknecht, R. & Jungbluth, N. (2007). Overview and Methodology, In: *Ecoinvent report No. 1*, Frischknecht, R. & Jungbluth, N. (Ed.), Dübendorf, Switzerland.

Garrido Fernandez, A. (1983). Study of ripe olive wastewaters and their reusing possibilities. *Grasas y Aceites*, 5, 317.

Garrido Fernàndez, A. & Fernàndez Diez, M. J. (1997). *Table olives, production and processing*. Chapman & Hall, London, UK, 1997.

Gomez, A. H. S.; Garcia, P. & Navarro, L. R. (2006). Elaboration of table olives. *Grasas y Aceites*, 57, 1, 86-94.

Hernandez, A.; Martin, A.; Aranda, E.; Perez-Nevado, F. & Cordoba, M. G. (2007). Identification and characterization of yeast isolated from the elaboration of seasoned green table olives. *Food Microbiology*, 24, 4, 346-351.

Higinio Sánchez Gómez, A.; García García, P. & Rejano Navarro, L. (2006). Elaboration of table olives. *Grasas y Aceites*, 57, 1, 86-94.

IKP, PE (2002). *GaBi 4 - Software-system and databases for life cycle engineering*. Stuttgart, Echterdingen. Germany.

ISO 14040. (2006). *Environmental management - Life cycle assessment - Principles and framework (ISO 14040:2006)*, International Standard, Geneva, ISO.

ISO 14044. (2006). *Environmental management - Life cycle assessment - Requirements and guidelines (ISO 14044:2006)*, International Standard. Geneva, ISO.

ISO/TS 14048. (2006). *Environmental management - Life cycle assessment – Data Documentation Format. (ISO/TS 14048: 2006)*, International Standard. Geneva, ISO.

Javier Casado, F; Antonio Higinio Sánchez, A.; Rejano, L. & Montaño, A. (2007). Estudio de nuevos procedimientos de elaboración de aceitunas verdes tratadas con álcali, no fermentadas, conservadas mediante tratamientos térmicos. *Grasas y Aceites*, 58, 3, 275-282.

Javier Casado, F.; Higinio Sánchez, A.; Rejano, L.; de Castro, A. & Montaño, A. (2010). Stability of sorbic and ascorbic acids in packed green table olives during long-term storage as affected by different packing conditions, and its influence on quality parameters. *Food Chemistry*, 122, 3, 812–818.

Kanavouras, A.; Gazouli, M.; Tzouvelekis Leonidas, L. & Petrakis, C. (2005). Evaluation of black table olives in different brines. *Grasas y Aceites*, 56, 2, 106-115.

Katsoni, A.; Frontistis, Z.; Xekoukoulotakis, N. P.; Diamadopoulos, E. & Mantzavinos, D. (2008). Wet air oxidation of table olive processing wastewater: determination of key operating parameters by factorial design. *Water research*, 42, 3591–3600.

Kotsou, M.; Kyriacou, A.; Lasaridi, K. & Pilidis, G. (2004). Integrated aerobic biological treatment and chemical oxidation with Fenton's reagent for the processing of green table olive wastewater. *Process Biochemistry*, 39, 11, 1653-1660.

Marsilio, V.; Campestre, C. & Lanza, B. (2001). Sugar and polyol compositions of some European olive fruit varieties (Olea europaea L.) suitable for table olive purposes. *Food Chemistry*, 72, 4, 485-490.

Marsilio, V. ; Russi, F. ; Iannucci, E. & Sabatini, N. (2008). Effects of alkali neutralization with CO_2 on fermentation, chemical parameters and sensory characteristics in Spanish-style green olives (Olea europaea L.). *LWT – Food Science and Technology*, 5, 796.

Notarnicola, B.; Tassielli, G. & Nicoletti, G. M. (2004). LCC and LCA of extra-virgin olive oil: organic vs. conventional. Life Cycle Assessment in the Agri-food sector, *Proceeding from the 4th International Conference "Life Cycle Assessment in the Agri-food sector:*, pp. 289-293, Horsens, Denmark, Danish Institute of Agricultural Sciences (DIAS), October 6-8 2003, Retrieved from
http://www.lcafood.dk/lca_conf/DJFrapport_paper_2_poster.pdf

Olivieri, G. ; Falconi, F. ; Pergreffi, R. ; Neri, P. & Romani, A. (2005). Life Cycle Assessment for environmental integrated system in the olive oil tuscan company. *European Meeting of the International Society for Industrial Ecology*, pp. 127-130, Bologna, Italy 10-11 January 2005, printed by Setac.

Olivieri, G. ; Neri, P. ; Bandini, F. & Romani, A. (2007). Life Cycle assessment of Tuscany olive oil production, *Proceedings of The AGS Annual Meeting*, Barcelona, Spain. 2007.

Quintana, M. C. D.; Barranco, C. R. & Garcia, P. G. (1997). Lactic acid bacteria in table olive fermentations. *Grasas y Aceites*, 48, 5, 297-311.

Raggi, A. ; Cichelli., A. ; Pagliuca, G. & Costantino, A. (2000). A screening LCA of olive husk combustion for residential heating, *1st World Conference on Biomass for Energy and Industry*, pp 1001-1004, Sevilla, Spain, 5-9 June 2000, James & James (Science Publishers) Ltd.

Romeo, F. V.; De Luca, S; Piscopo, A.; Perri, E. & Poiana, M. (2009). Effects of post-fermentation processing on the stabilisation of naturally fermented green table olives (cv. Nocellara etnea). *Food Chemistry*, 116, 4, 873–878.

Russo, C. ; Cappelletti, G. M. & Nicoletti, G. M. (2008). LCA of Energy Recovery of the Solid Waste of the Olive Oil Industries, *6th International Conference on Life Cycle Assessment in the Agri-Food Sector "Towards a Sustainable Management of the Food Chain"* Zurich, Switzerland, November 12–14, 2008, Retrieved from: http://www.agroscope.admin.ch/aktuell/02720/02722/03985/index.html?lang=de

Russo, C. ; Cappelletti, G. M. & Nicoletti, G. M. (2010). Life Cycle Assessment (LCA) used to compare two different methods of processing ripe table olives. *Grasas y Aceites*, 61, 2, 136-142.

Salomone, R. ; Cappelletti, G. M. ; Ioppolo, G. ; Mistretta, M. ; Nicoletti, G. M. ; Notarnicola, B. ; Olivieri, G. ; Pattara, C. ; Russo, C. & Scimia, E. (2010). Italian experiences in Life Cycle Assessment of olive oil: a survey and critical review, *Proceeding of the 7th International Conference on Life Cycle Assessment in the Agri-Food Sector"*, pp. 265:270, Bari, Italy, September 22–24, 2010.

Segovia-Bravo, K. A.; Garcia-Garcia, P.; Arroyo-Lòpez, F. N.; Lòpez-Lòpez, A. & Garrido-Fernàndez, A. (2008). Ozonation process for the regeneration and recycling of Spanish green table olive fermentation brines. *Eur Food Res Technol*, 227, 463–472.

Unal, K. & Nergiz, C. (2003). The effect of table olive preparing methods and storage on the composition and nutritive value of olives. *Grasas y Aceites*, 54, 1, 71-76. www.internationaloliveoil.org (Website consulted Sep 2010).

10

Effects of Wastewater Treatment Plant on Water Column and Sediment Quality in Izmir Bay (Eastearn Aegean Sea)

F. Sanem Sunlu, Ugur Sunlu, Baha Buyukisik,
Serkan Kukrer and Mehmet Aksu
Ege University, Faculty of Fisheries, Dept. of Hydrobiology
Turkey

1. Introduction

Economic and social consequences of damage to the marine environment are becoming increasingly evident. Unless seas and oceans are carefully protected, their economic potential can not be sustainable. The marine environment is one of humanity's most precious assets. Oceans and seas cover 71% of the earth's surface and are the greatest sources of biodiversity, containing 90% of the biosphere. Marine ecosystems play a key role in climate and weather patterns. They also contribute to economic prosperity, social well-being and quality of life and are literally a source of survival for coastal communities. However, this environment is under intense pressure. The pace of degradation of biodiversity and habitats; the level of contamination by dangerous substances and the emerging consequences of climate change are some of the most visible warning signals (Environment for Europeans 2005). Only recently marine eutrophication is being regarded as pollution, particularly in near shore environments where more often low water transparency, oxygen depletion and algal blooms occur. Nutrient concentrations in sea water and sediment increase remarkably going from offshore to inshore, due to the proximity of terrestrial and domestic inputs and to the increase of biotic and abiotic processes strictly related to the progressive decrease of water depth.

The Bay of Izmir is in a state of pollution centre in Turkish Aegean coast region in respect of aesthetic and welfare where pollution increased in the course of time from what it used to be in 1960s. The most important factors of this current status are; domestic wastes of more than 3 million people; industrial wastes from 1,500 factories; wastewater discharge during maritime transportation and shipyard services filling materials arisen from the recreation of seaside alluvions carried with rivers and valleys. Izmir Bay is surrounded by major agricultural plateau. Menemen plateau in the North–North West of Izmir is one of the most important production fields where agricultural irrigation is utilized. The Bay is also influenced by the pollution caused by the agricultural activities in the Gediz River water shed and erosion of a large area by Gediz River.

The bay of Izmir, which is the biggest harbour on the Aegean Sea, is of economical importance for Izmir, the third largest city in Turkey. The Bay is divided into inner, middle and outer bays in terms of topographical and hydrographical characteristics. The inner bay is considerably small in area (57 km²) and shallow in depth (max. 15 m). It had received the majority of domestic and industrial wastewaters before the construction of wastewater treatment plants. This section of the bay still receives some inflow of fresh water from several creeks which are mostly polluted by industrial wastewaters.

Because of limited water exchange with the Outer Bay and Aegean Sea, pollution of the Inner Bay had reached unacceptable levels. Eutrophication of the Inner and the Middle Bay had started and spread progressively to the outer part of the Bay. Red-tide occurrence was reported to have increase in frequency in last decade (Sunlu et al. 2007). For this reason Izmir Municipality decided to construct Izmir Big Channel WasteWater Project in 1969. However, wastewater treatment plant construction completed in 2002. At the end of the plant construction, the pollutant levels of the Inner Bay water decreased slowly and recovery period has begun (Kaymakci et al. 2000). This is why, the pollutant levels of the Inner Bay water decreased slowly.

The aim of this research is to determine the effects of "Izmir Big Channel Wastewater Treatment Project" to the sediment quality and water column of Izmir Bay. For this purpose, seawater samples and sediment samples were collected from three stations which are located in the middle and inner parts of the Izmir Bay. The water samples were collected as a weekly and sediment samples as monthly intervals during 2003.

2. Materials and methods

2.1 Location and sampling

In this study, three stations were chosen for sampling, two in the inner and one in the middle part of the Izmir Bay (Figure 1). Station 1 is near Izmir Harbor (Inner Bay) 38°27'17"N-27°09' 37"E. Station 2 is offshore of the Karsiyaka Yacht Club (Inner Bay) 38°26'86"N-27°06'56"E and Station 3 is offshore of the Wastewater Treatment Plant (Cigli, Middle Bay) 38°25'47"N-27°00'05"E.

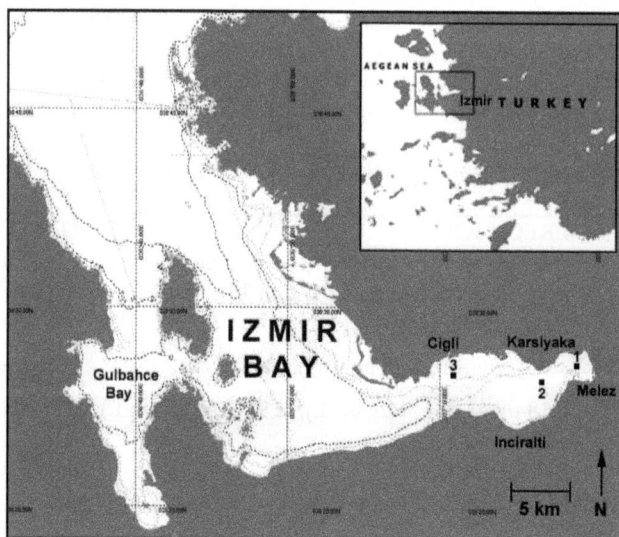

Fig. 1. Map of Izmir Bay and the sampling stations

The sampling points are shown in Fig. 1. The first sampling station is located in the area of the outflows of the Melez stream and consequently receives agricultural, domestic and industrial discharges carried on by it. This station is also influenced by the harbour activities of İzmir. The second station in Karşıyaka, shows characteristics of mixture of ST 3 and ST 1 according to current system of the İzmir Bay. The last station in Cigli where the physical and biological waste water treatment plant exists is affected by the current system of the bay. It was particularly chosen to better understand effects of this plant on İzmir Bay.

Physico-chemical environmental parameters, nutrients and some general biological parameters, were measured weekly during one year period. All these parameters were measured at different depths of the three selected sampling stations.

For this study water samples were collected using a peristaltic pump and screened 280 μ mesh to remove macrozooplankton. Polycarbonate bottles of 20 L capacity were filled with sea water and moved to the laboratory.

2.2 In situ measurements

Seawater temperature was recorded by an electronic thermometer with a sensitivity of ±0.1 °C. The pH of the samples were also measured on-site using a pH-meter (Hanna Ins.). Likewise, the dissolved oxygen concentration (DO) was measured with a portable dissolved oxygen-meter (YSI, Model 55).

2.3 Analytical measurements

The salinity of the seawater was determined by the Harvey method. The samples collected from the three different stations and different water depths were kept in 1 L polyethylene bottles and analyzed for nitrate (NO_3^-), nitrite (NO_2^-), ammonium (NH_4^+), silicate and Reactive Phosphate (RP) using the methods by Strickland and Parsons 1972; Wood 1975; Parsons et al. 1984.

For chlorophyll a and phaeopigments, given amount of surface seawater was filtered through GF/C filterpaper using the Milipore Filtration system. The analyses were carried using a UVD spektrophotometer (Bosch-Lomb Spectronic 21) according to the method by Strickland and Parsons (1972). Particulate organic carbon (POC) analyses were carried out using wet oxidation method and spectrophotometry (Strickland and Parsons 1972; Parsons et al. 1984).

The detection limits and precision of methods used were given in Table 1.

PARAMETER	PRECISION	DETECTION LIMITS
NO_3^-	± 0.2 μg atN/L (1 μg atN/L at 1.7 cm. cell)	0.1-45 μg at/L
NO_2^-	± 0.2 μg atN/L (1 μg atN/L at 1.7 cm. cell)	0.1-2.5 μg at/L
NH_4^+	± 0.7 μg atN/L (10 μg atN/L at 1.7 cm. cell)	0.2-10 μg at/L
Reactive Phosphate (RP)	± 0.03 μg atP/L (3 μg atN/L at 1.7 cm. cell)	0.05-5 μg at/L
Si	0,18 μg at/L (~4 μg at/L); ± 9 (~150 μg at/L)	0,26-400 μg at/L
Salinity	± 0.05 psu	
Chl a	± 0.2 μg atN/L	0.2-50 μg Chl a,b,c /L
Dissolved Oxygen (DO)	± 0.3 mg/l (± 0.2 °C)	

Table 1. The detection limits and precision of methods used.

Sediment samples were collected from these three stations on a monthly basis between January and 2003 December 2003. Sediment samples were collected using Van-Veen Grap. Chlorophyll Degradation Products (CDP) were analyzed through acetone extraction and spectrophotometry (Lorenzen 1971). Organic carbon values were determined according to Modified Wakley-Black Titration Method (Gaudette et al. 1974).

3. Results

3.1 Water column

In this study, some environmentally important parameters and nutrients were measured weekly during one year period at different depths of 3 selected sampling stations in Izmir Bay.

Table 2 gives the minimum, maximum and average (± standart errors) values of the physico-chemical parameters related to the water samples from the Izmir Bay.

	Station 1		Station 2		Station 3	
	Range	Mean±SE	Range	Mean±SE	Range	Mean±SE
Temperature	9.0-28.2	18.68±0.5	8.9-27.4	18.56±0.42	9.6-28.0	18.90±0.37
Salinity	31.93-43.85	39.84±0.13	33.97-43.85	39.98±0.11	33.97-44.85	39.9±0.1
pH	7.4-8.6	8.03±0.01	7.5-8.7	8.08±0.01	7.5-8.6	8.09±0.01
DO	3.86-14.40	7.44±0.14	4.57-13.60	7.72±0.12	4.16-12.9	7.72±0.09
NH_4-N	0.21-36.97	7.83±0.56	0.00-32.19	4.89±0.30	0.09-40.94	3.84±0.29
NO_3-N	0.00-19.31	4.55±0.38	0.00-21.35	3.50±0.26	0.00-17.63	2.10±0.16
NO_2-N	0.00-28.99	3.54±0.38	0.00-16.99	2.54±0.24	0.00-9.69	1.06±0.10
PO_4-P	0.60-16.05	3.67±0.16	0.54-19.56	3.51±0.18	0.00-31.43	2.77±0.21
Si	0.31-43.89	12.62±0.77	0.47-54.12	11.47±0.64	0.16-41.80	8.81±0.54
N/P	0.57-15.69	5.43±0.29	0.23-20.52	4.46±0.26	0.00-53.65	4.36±0.40
Si/P	0.22-28.03	4.31±0.35	0.27-56.38	4.56±0.40	0.00-83.38	5.60±0.61
Chl a	0.00-66.13	5.72±0.59	0.00-23.55	4.65±0.28	0.00-12.82	2.78±0.17

Table 2. The minimum, maximum, average and standart errors of the physico-chemical parameters related to the water samples from the Izmir Bay. Temperature (°C), salinity (%o), DO (mg/l), NH_4^+-N NO_3^- -N, NO_2^--N, PO_4^{-3}-P, Si (µM), Chl a(µg/l).

3.1.1 Physico-chemical parameters

Rainfall, evoporation, streams and wastewater discharge effect variations of salinity in the bay of İzmir. As a result of rain in winter salinity has decreased. In addition salinity began to increase in during spring months suggest that water discharge out of the treatment plant left the bay along the northern shores.

Considerable increases in dissolved oxygen concentrations were observed in the first week of january, the second of march, the third of april, the first of may and the third of june. Although fluctuations were found between january and july, there was a general decrease in the values concerned, which seems consistent with the general increase found in water temperature, suggesting that dissolution of gases in water diminishes based on temperatural increase. A very slow increase was observed in dissolved oxygene concentrations from early july to the end of the year. Dissolved oxygen concentrations at ST 1 and ST 2 were found to be below the oxygen saturation.

Increases seen in pH, one of physico-chemical environmental parameters between the first week and 4th week of May were associated with those in chlorophyll a values, suggesting a decrease in inorganic carbon induced by photosynthesis, which is believed to have been caused by any indirect biological or chemical phenomenon. While significant decreases in pH appeared in mid of august, end of september and november, the general tendency to diminish in pH seems consistent with drop of dissolved oxygen saturation.

3.1.2 Nutrients

One of parameter groups quite influencing lower trophic levels in aquatic ecosystems is nutrients. Accordingly, variable values of nitrogen forms, orthophosphate and silicate based on time and depth at the three stations chosen in Izmir bay are as follows.

When nutrients and chlorophyll-a concentrations were compared to the studies carried out before the construction of Wastewater Treatment Plant (WWTP), significant decreases were observed for the nutrients (Table 6), but chlorophyll-a concentrations were higher than the values determined after WWTP by Kukrer and Aydin (2006). This situation points out the role of primary production on reduction of nutrient concentration; and thus, it is thought that this reduction transformed into the phytoplankton biomass (Kukrer, 2009).

A similarity in the spatio-temporal distributions of NH_4-N and NO_2-N were observed at all stations. Bizsel and Uslu (2000) explain this similarity with nitrification process: NH_4-N rapidly transforms into NO_2-N but transformation of NO_2-N to NO_3-N is a slower process (Ozkan et al., 2008). Hence, NH_4-N and NO_2-N had similar trend in this study. NO_2-N values were lower than NO_3-N over the sampling period. Morris (1980) was reported that this situation was normal and NO_2-N accumulates distinguishable under low DO condition (Kukrer, 2009).

Koray et al. (1992) emphasized that a big part of total nitrogen in the polluted Izmir Bay was the ammonium from industrial and domestic wastes. Contrary, in this study nitrate have the biggest share in total nitrogen concentration due to the wastewater treatment plant which reduces ammonium inputs. Additionally, ammonium concentration is kept under control by phytoplankton over a year. In spite of this progress, the ammonium enrichment continues owing to the creeks and sediment which have high ammonium concentration (Ozkan et al., 2008).

During the phytoplankton bloom period (January-August), NH_4-N, $[Si(OH)_4$-Si] and o.PO_4-P concentrations were lower than the values in Autumn. While inverse trend between NH_4-N and chlorophyll-a shows consumption of ammonium by phytoplankton, a similar relationship could not be observed between NO_3-N and chlorophyll-a (Kukrer, 2009).

The opposite trend between NH_4-N and chlorophyll-a showed the consumption of ammonium by phytoplankton, but an expected negative correlation between NO_3-N and chlorophyll-a could not be found. It can be explained that NH_4-N could block the uptake of NO_3-N and/or NH_4-N might be preferred by phytoplankton (Morris, 1980).

The study in which data was processed into Principal Component Analysis (PCA) aimed at determining contributions of the measured variables to total variance with different analysis for each station. All variables were defined as four major components at ST 1, explaining for 64.5% of total variance. Thus, temperature, phosphate, pH and phaeopigment account for 22.1 % of the variations involved. 18.4 % of them is to a great extent governed by DO, chl a, salinity and nitrite. 12.6 % is mostly controlled by silicate and ammonium whereas 11.3 % generally by nitrate (Table 3).

ST 1	Component 1	Component 2	Component 3	Component 4
Phaopigment	**0,332992**	0,00587052	0,0782505	-0,345222
Temperature	0,148211	**-0,422099**	-0,15314	-0,234076
Salinity	**0,574924**	-0,113647	-0,0567523	-0,00371232
pH	**0,381094**	0,334592	-0,0951191	0,0634268
PO$_4$	**0,472503**	-0,0803457	0,0848547	0,368648
NO$_3$	-0,166612	-0,214467	-0,196831	**0,691634**
NO$_2$	-0,106301	**-0,364725**	-0,157476	0,155942
NH$_4$	0,0223274	0,240198	**0,656172**	0,133631
SiO$_4$	-0,0478669	-0,198481	**0,617408**	0,186501
DO	-0,292287	**0,478749**	-0,134424	-0,066133
Chl -a	0,20209	**0,430308**	-0,239983	0,355532

Table 3. Component Weights of ST 1

All variables of PCA made for ST2 were defined as four major components at ST 1 which explained for 62% of total variance. 21% of it often depends on temperature, phosphate and DO whereas its 17.4 % is mostly controlled by nitrite, nitrate, pH. 14% of variance is found to be accounted for by ammonium and silicate while its 9.6% is to a great measure under the control of phaeopigment, salinity and chl a (Table 4).

ST 2	Component 1	Component 2	Component 3	Component 4
Phaopigment	0,190085	0,033317	-0,337102	**0,416977**
Temperature	**0,612123**	-0,0620875	0,00251255	0,050634
Salinity	0,175982	-0,389765	-0,0522611	**0,45086**
pH	0,350358	**0,409077**	-0,0404681	-0,241227
PO$_4$	**0,443203**	0,0770914	-0,0509096	-0,124404
NO$_3$	-0,120681	**-0,436475**	-0,160533	-0,282205
NO$_2$	-0,0501546	**-0,528066**	-0,152599	0,0743837
NH$_4$	-0,0090085	0,17262	**0,513087**	0,382987
SiO$_4$	0,0249009	-0,156039	**0,622841**	0,218553
DO	**-0,469988**	0,328992	-0,165754	0,194231
Chl -a	0,0188411	0,195156	-0,389454	**0,478833**

Table 4. Component Weights of ST 2

All variables were described as four major components at ST 3 which explained for 61.2% of total variance. 25.2 % of total variance is generally explained by temperature, phosphate, oxygen and phaeopigment Nitrate is seen to be responsible for 14.9 % of it whereas its 11.8% is basically governed by salinity, chlorophyll a and nitrite. On the other hand 9.3 % of total variation is mostly controlled by silicate and ammonium (Table 5).

ST 3	Component 1	Component 2	Component 3	Component 4
Phaopigment	**0,309441**	-0,115255	-0,118439	0,00891844
Temperature	**0,478778**	0,368824	-0,00879898	0,140916
Salinity	-0,024596	0,365174	**-0,519479**	0,253368
pH	**0,33159**	0,186109	0,321044	-0,0842017
PO$_4$	**0,437231**	-0,22811	0,0675536	0,134745
NO$_3$	0,200952	**-0,554785**	-0,129602	0,162289
NO$_2$	0,0991039	-0,401237	**-0,46215**	0,29997
NH$_4$	0,285814	-0,197389	0,0871168	**-0,468515**
SiO$_4$	0,244234	-0,0379852	-0,22798	**-0,566497**
DO	**-0,383154**	-0,345061	0,258506	-0,0662661
Chl -a	0,186924	-0,0491215	**0,501633**	0,479047

Table 5. Component Weights of ST 3

Table 6 shows minimum and maximum values of nutrients and *Chl a* in some previous studies which were carried out in the different parts of the Izmir bay. Izmir Wastewater Treatment Plant Construction was completed in the 2002. It works on the principle of nitrogen and phosphorus treatment technology with activated sludge. Previous studies indicated that the concentration of TNO$_x$-N has been reduced during after wastewater activated sludge technology plant except sudden discharge, while reactive phosphate concentrations were increased in the Bay. In the Middle and Inner Parts of the Bay Chlorophyll *a* concentration has been gradually reduced after treatment.

In conclusion, we are of the opinion that it would be of great use to develop and plan further similar studies periodically and for the long run considering that they could shed light on precautions to be taken in terms of both environmental and public health.

The changes in the state variables of ecological model for İzmir Bay before and after the sewage treatment has been given by Büyükışık et al., 1997 (Fig.2 and 3). They reported that average light intensities in water column would be recovered in a year if the treatment plant begins to work. Indeed, after one year from starting of sewage treatment (2003), the observation in recovery of the average light intensities in water column consistent with the model outputs in case of treatment.

But some changes in temporal variations of phytoplankton biomass has been observed (Fig.4). Some exceptional blooms has taken place in mid-winter, early summer and autumn. Model does not includes the kinetic parameters of *Ditylum brightwellii* (in winter) and *Rhizosolenia setigera* (in summer).

These two species are relatively large sized phytoplankton and they contributed greatly to the total phytoplankton carbon and POC values.

Specially some members of genus Rhizosolenia can change their cellular density, sink deeper, uptake and storage the nutrients and go on their growth.

Table 6. Minimum and maximum concentrations of nutrient and chlorophyll-a in Izmir Bay and Aegean Sea from different studies

Locations	Period	NO$_3$(μM)	NO$_2$(μM)	NH$_4$(μM)	Si(μM)	RP(μM)	Chl a(μg l^{-1})	Reference
Inner part of Izmir Bay	1993-1994	BDL-3,04*	BDL-4,65*	0,12-468*	-	0,36-49	BDL-189	Bizsel, Uslu,2000
Middle part of Izmir Bay	1993-1994	BDL-3,49	BDL-3,57	BDL-44	-	0,06-3,79	0,5-62	Bizsel, Uslu,2000
Outer part of Izmir Bay	1993-1994	BDL-4,91	BDL-0,16	BDL-11,11		BDL-6,42	BDL-2,95	Bizsel, Uslu,2000
Inner part of Izmir Bay	1993-1994	BDL-3,11	BDL-4,65	BDL-468	-	0,18-49	BDL-189***	Bizsel, Uslu,2000
Candarlş Bay (Aegean Sea)	1994-1995	0,001-0,31	BDL-0,1	0,42-2,38	27,74-63,19	BDL-0,48	BDL-1,13	Aksu et.al.2010
Middle-Inner part of Izmir Bay	1996-1998	0,13-27	0,01-18	0,10-21	0,50-39	0,01-10	0,10-26	Kucuksezgin, et. al. 2006
Middle-Inner part of Izmir Bay	2000	0,15-18	0,02-12	0,13-34	0,43-20	0,13-3,8	0,46-18	Kucuksezgin, et. al. 2006
Middle-Inner part of Izmir Bay	2001	0,29-16	0,02-4,3	0,11-50	1,2-18	0,14-2,9	0,38-7,8	Kucuksezgin, et. al. 2006
Cerence Bay (Aegean Sea)	2002	0,26-6,7	0,01-6,1	0,10-6,7	1,0-26	0,14-4,4	0,13-3,7	Colak-Sabancş Koray, 2001
Middle-Inner part of Izmir Bay	2002	0,04-2,19	BDL-2,51	BDL-3,53	-	BDL-2,82	BDL-0,320	Aydşn Gençay, Büyükşşk, 2006
Middle-Inner part of Izmir Bay	2003	0,12-8,6	0,01-1,0	0,12-2,4	2,6-32	0,32-4,5	0,24-2,6	Colak-Sabancş Koray, 2001
Inner part of Izmir Bay	2007-2008	1,54-11,77	0,00-3,51	0,23-22,28	1,99-41,94	0,00-5,96	5,03-30,26	Kukrer, 2009
This Study	2003	BDL-21,35	BDL-28,99	BDL-40,94	0,16-54,12	BDL-31,43	BDL-66,13	This Study

* Min-Max;
** Average value;
*** Data from (32);
BDL: Below Detection Limits

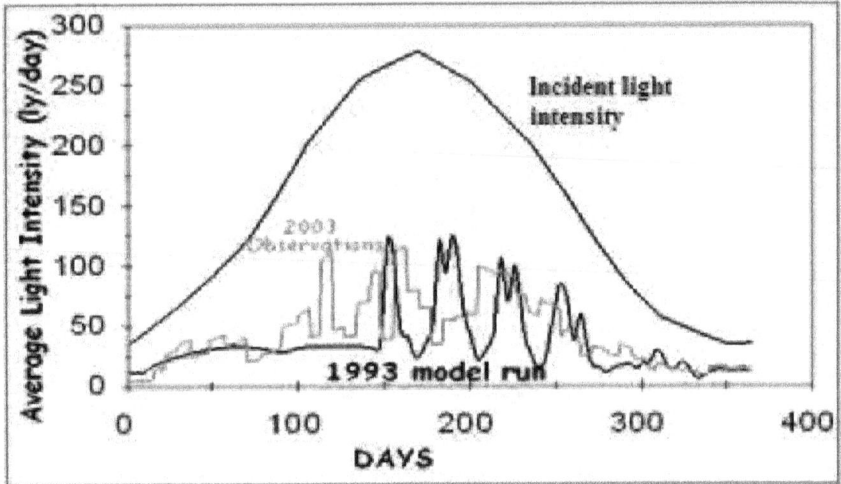

Fig. 2. Temporal changes of the average water column light intensities obtained from model in 1993 (Black curve, Büyükışık et al 1997) and from chl-*a* values in 2003 (gray lines, Sunlu et.al, 2007). The black curve at top represents the temporal changes in incoming sub-surface light intensities (Büyükışık et al 1997).

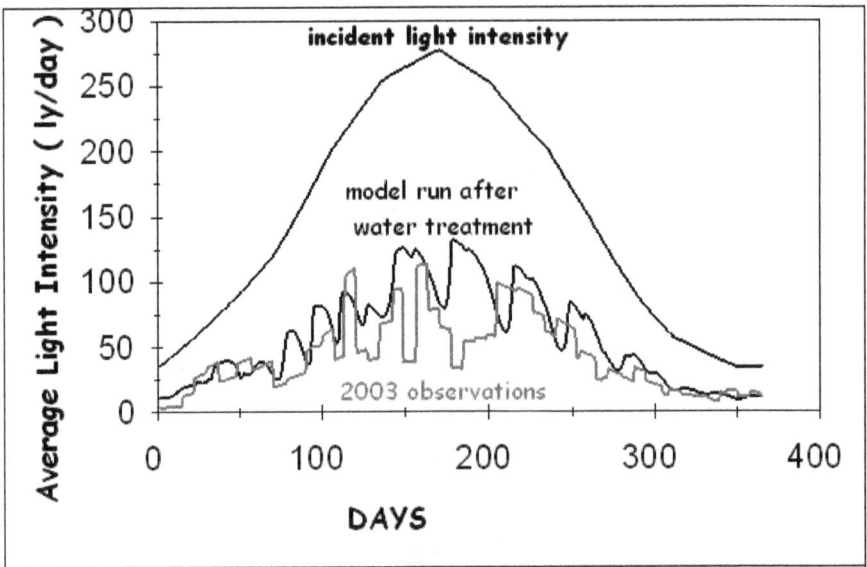

Fig. 3. Temporal changes of the average light intensities obtained from model in case of 90% nutrient treatment (black curve, Büyükışık et al 1997) and from chl-*a* values in 2003(gray lines, Sunlu et al, 2007). The black curve at top represents the temporal changes in incoming sub-surface light intensities (Büyükışık et al 1997).

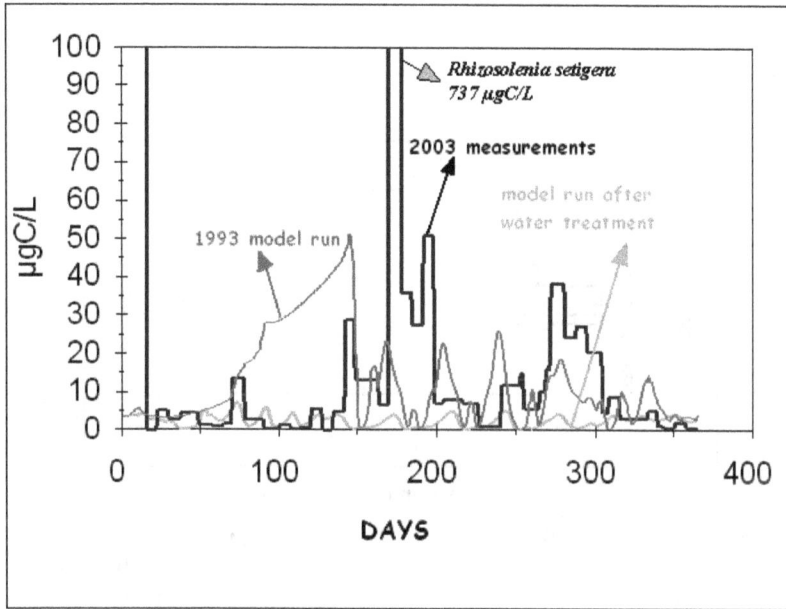

Fig. 4. Temporal changes of the phytoplankton biomass obtained from model in case of 90% efficiently treatment (light gray curve, Büyükışık et al 1997). The dark gray curve represents the model run outs in 1993 (moving average, Büyükışık et al 1997). Black column in graph represents the measurements in 2003 from biomass calculates two microscopic examinations (Sunlu et al, 2007).

3.2 Sediment

Values measured at stations ranged between; 0.09–9.32 µg/L for phaeopigment, 0.05–1.91 mg/L for particulate organic carbon in sea waters, 11.88–100.29 µg/g for chlorophyll degradation products and 1.12–5.39% for organic carbon in sediment samples. In conclusion, it was found that grazing activity explained carbon variations in sediment at station 2, but at station 1 and station 3 carbon variations in sediment were not related to autochthonous biological processes.

3.2.1 Organic carbon in sediment

Organic carbon values at station 1 ranged from 2.63 to 3.39%. Average concentration was 3.03%. Minimum, maximum and average organic carbon values at station 2 were 1.73, 5.39 and 4.33% respectively. Organic carbon values at station 3 ranged from 1.12 to 2.41%. Average concentration was 1.58% (Fig. 5). Previous carbon contents in the sediment samples from the different regions of Aegean Sea were given in Table 7.

3.2.2 Chlorophyll degradation products in sediment (CDP)

Chlorophyll degradation products in sediment at station 1 ranged from 50.79 to 90.66 µg/g and average value was found 62.62 µg/g. At station 2 average CDP value was 81.39 µg/g. Minimum and maximum values were measured as 41.58–100.29 µg/g respectively. CDP

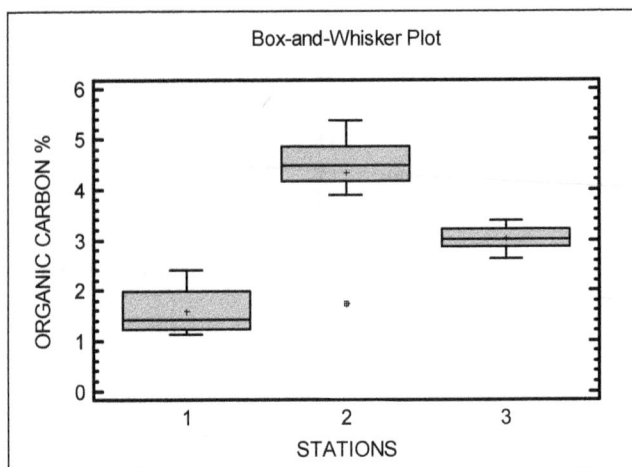

Fig. 5. Box and whisker plot of Organic carbon (%) values at all sampling stations.

concentrations at station 3 ranged from 11.88 to 52.12 µg/g. Annual mean was 34.44 µg/g(Fig. 6). When each three region was discussed separately, at the Station 2, algal sedimentation and/or mesozooplankton grazing explain variations of carbon in the the sediment samples (r=0.7879 p=0.0023). According to statictical analyses of C sed/CDP for each region, variations of CDP in sediment seems independent from carbon in sediment variations for station 1 and station 3 in sequence (r=0.339, r=0.206). Melez, Manda and Arap Rivers discharge their waters rich in organic mater around station 1 (Turkman 1981). At station 3, during the year CDP concentrations were at the lowest value and it can be explained by background carbon levels that mask carbon variations which is caused by algae (< %2). Besides, the output of the wastewater treatment plant is close to the station 3 and it constitutes crucial silicate source. Diatoms consist of skeleton with silica are known as having five times lower carbon content than Dinoflagellates (Hitchcock 1982 in Smayda 1997). That situation can explain that during the year phytoplankton community has lower carbon content. Even if export production to sediment increases relatively low productivity and low carbon content in water column can cause a similar situation in diatom dominated marine environments. By using overall data in Inner and Middle Izmir Bay, chlorophyll degradation products in sediment versus carbon values were plotted. A good linear relationship between CDP and carbon was obtained (r2=0.771, p=0.000):

$$[Carbon]_{sed}=0.2077+0.0466*[CDP]_{sed}$$

A general equation was found for predicting the Izmir Inner Bay's CDP and organic carbon values in sediment. It was found that there are no significant differences in sediment carbon values depending on time but spatial variations related to sampling stations are more evident. When spatial scale is widened, CDP variations explained 77% of carbon variations in the sediment for overall data. Approximately 23% of these variations were originated from allochthonous sources.

At station 3, it is possible that grazing on diatoms and/or mixotrophy in dinoflagellates are dominant on certain onths of the year. Consequently, it is not possible to explain variations of the carbon in sediment with the pigment contents of sediment. Station 2 has highest carbon and CDP values and also has a relationship between CDP and organic carbon content. This situation can be explained by the fact that station 2 is relatively away from external sources and has high biological activity (Sunlu et al. 2007). At station 1, however, relation is weak despite higher carbon and CDP values than at station 3. Contribution of external carbon sources as rivers may play important role on this weak correlation.

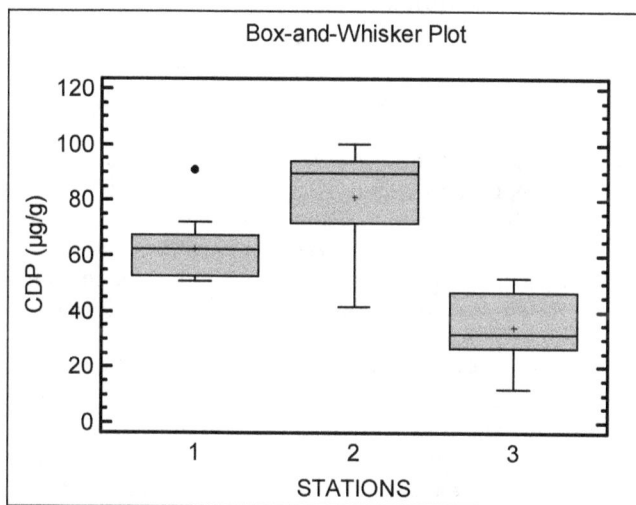

Fig. 6. Box and whisker plot of CDP (µg/g dry sediment) values at all sampling stations.

Locations	Carbon in Sediment (%)	Reference
Middle part of Izmir Bay	0.87-1.60	Yaramaz et. al. 1992
Inner part of Izmir Bay	0.57-3.42	Yaramaz et. al. 1991
Izmir Bay	11.4	Anonymous, 1992
Izmir Bay	2.0-7.0	Anonymous, 1997
Gulluk Bay (Southern Aegean Sea)	0.1-4.5	Egemen et. al. 1999
Gulluk Bay (Southern Aegean Sea)	1.07-2.13	Atılgan, 1997
Urla (Middle part of Izmir Bay)	1.25-2.1	Sunlu et. al. 1999
Pariakos Bay (Greece)	0.15-11.01	Varnavas and Ferentionos, 1982
Evoikos Bay (Greece)	1.2	Scoullos and Dassennakis, 1982
Evoikos Bay (Greece)	0.66-2.4	Angelidis et. al. 1980
Southern Turkish Aegean Sea	1.3-13.1	Aydın and Sunlu, 2005
Northern Turkish Aegean Sea	0.35-15.63	Sunlu et. al. 2005
Middle part of Izmir Bay	1.12-5.39	This Study

Table 7. Previous carbon contents in the sediment samples from the different regions of Aegean Sea.

4. Conclusion

When our mean results were compared with those obtained before Izmir wastewater treatment plant was operating, concentrations of chlorophyll a and nitrogen forms declined while it was not the case for orthophosphate.

The fact that the processes affecting Reactive Phosphate (RP) and TIN occur at different times indicates important differentiations in the temporal variations of these two nutrients in the Inner Bay. From the distribution of the nutrients and their percentages, important evidence regarding the process have been gathered. These processes:

- Inflow with the creeks is especially evident during rainfall and there is a big increase in Si and Nitrogen forms.
- Rapid decreases of freshwater inflows from rainfall based on current global warming tend to restrict Si and N inflows. Water outflow treated from treatment plant is another source of nutrient with N/P ratios being about <=2. RP induced by water from treatment plant thus contributes to RP reserves in Inner Bay.
- The winds, although increasing fresh water inflow and water column, frequently carry the deep water to the surface. This shows that the Inner Bay is often subject to a deep-water-based nutrient enrichment.

The phytoplankton blooms caused by the inflow of nutrients to the Inner Bay in turn result in the intake of nutrients by the phytoplanktons (especially diatoms) which are then exported to the deep waters and constitute the fuel for future phytoplankton blooms. Thus, the horizontal exportation of the nutrients out of the Inner Bay remains limited. It is only due to the winds that the wastewaters flow outwards from time to time.

Because total renewal of Inner Bay water by the current system takes about ten days, nutrient load provided by various sources in the area is most important reason for overgrowth of phytoplanktons observed in the Izmir Bay.

Silicate is essential for the diatoms to compete effectively with dynophylagellates and plays an important role in the increase in species in the bay and this nutrient, coming with the rainfall from the shore in non-point sources and point sources (i.e. creek, river), is of great importance for the Inner Bay.

We believe that unless the nutrient levels in the rivers are decreased, the Bay will continue its current state for a long time. Although a decrease has been observed in the nitrogen nutrients after the start of the wastewater treatment plant, former studies have shown that the phosphate concentrations have not changed and that the plant has been ineffective regarding this subject. The effective treatment of phosphate will be an important precaution against the new strategy that the phytoplankton might take up against the decreasing TIN.

The reason for this was that 2- 10 years elapsed between the two studies and the treatment facility begun to work in full capacity in 2002. On the other hand; carbon contents in the sediment samples of our study are considerably lower compared with the values obtained in a large scale previous research carried out by different regions around Aegean Sea.

General sediment texture of Izmir Bay was studied by Duman et al. (2004). Average sediment particle size was reported to be 4-8 φ and sediment texture to be sandy-silt. In Izmir Bay sorting coefficient indicates very poorly sorted deposits (SD=2-3). Prevailing wind direction in inner part of Izmir Bay was noted as Western and it has been reported that deep flow was toward to East and surface flow toward to West. Most of organic material remains in the silt near the pollution source and the correlation between grain size fractions and organic carbon was found to be highest in silt (Duman et al. 2004). One sediment component, vermiculite was found in the inner part of Izmir Bay at a rate of 3-11% and its main source was from Melez River (near station 1). Caolinit was found at a rate of 8–12% with neogen sediments coming from the rocks around the Bay (Aksu et al. 1998). Percentage of organic carbon was reported to be between 0.40 and 5.39 by Duman et al., from Izmir Bay (Duman et al. 2004). Range for these values was found to be between 1.12 and 5.39% in our study. These values were higher than previous report (Duman et al. 2004). The reason for this was that 2– 10 years elapsed between the two studies and the treatment facility begun to work in full capacity in 2002. On the other hand; carbon contents in the sediment samples of our study are considerably lower compared with the values obtained in a large scale previous research carried out by different regions around Aegean Sea (Table 7). It can be said that high carbon levels observed in inner part of Izmir Bay were from raw sewage and industrial outfalls carried by Melez River at station 1. But at station 2 and 3 high carbon levels were due to organic material formed by secondary pollution. The biggest contribution to the sediment is provided byexport production which was especially effective at station 2. A general equation was found for predicting the Izmir Inner Bay's CDP and organic carbon values in sediment. There are no significant differences in sediment carbon values depending on time but spatial variations (related to sampling stations) are more evident . In conclusion, it was found that carbon variations in sediment at station 2 (Karşıyaka, Offshore of the Yatch Club) can be explained by grazing activity, but at station 1 (Melez, Izmir Harbour) and station 3 (Cigli, Offshore of the Wastewater Treatment Plant) carbon variations in sediment could be related not only with autochthonous biological processes but also with physical processes (e.g. sweeping out of plant material by advection from the Bay). Especially wastewater treatment improves the water quality, but sediment does not respond to this treatment as fast as water column. Improvement in the quality of bottom water and sediment is the evidence of the recovery of the whole ecosystem of the Izmir Bay.

In conclusion, it was found that carbon variations in sediment at station 2 (Karşıyaka, Offshore of the Yatch Club) can be explained by grazing activity, but at station 1 (Melez, Izmir Harbour) and station 3 (Cigli, Offshore of the Wastewater Treatment Plant) carbon variations in sediment could be related not only with autochthonous biological processes but also with physical processes (e.g. sweeping out of plant material by advection from the Bay).

Especially wastewater treatment improves the water quality, but sediment does not respond to this treatment as fast as water column. Improvement in the quality of bottom water and sediment is the evidence of the recovery of the whole ecosystem of the Izmir Bay.

5. Acknowledgments

The authors would like to thank TUBITAK (Turkish Scientific and Technical Research Council) Project no: 102Y116, Izmir Municipality Gulf Control Staff and Science and Technology Research Centre of Ege University (EBILTEM) for their efforts to join of this project and their scientific and financial supports.

6. References

Aksu, A. E., Yatar, D., & Uslu, O. (1998). Assessment of marine pollution in Izmir Bay; heavy metal and organic compound concentrations in surficial sediments. Turkish Journal of Engineering and Environmental Science, 22, 387–415.

Aksu, M., Kaymakçı Basaran, A. & Egemen,Ö. (2010). Long-term monitoring of the impact of a capture-based bluefin tuna aquaculture on water column nutreint levels in the

Eastern Aegean Sea, *Environ Monit Assess*, 171: 681-688, DOI 10.1007/s10661-010-1313-y.

Angeldis, M., Grimanis, A. D., Zafiropoulos, D., & Vassilaki- Grimani, M. (1980). Trace elements in sediments of Evoikos Gulf, Greece. *Ves Journe'es Etud. Poll.Cagliari, CIESM*, 413–418.

Anon (1992). *Environmental impact report of alternative dumping areas of Izmir Harbour and approaching channeldredged material, (in Turkish)*, Dokuz Eylul Universitesi Deniz Bilimleri Enst. s, 14–15. Anon (1997). Izmir Bay Marine Research 1994–1998 (1997 Report), (in Turkish), D. E. U. Den. Bil.Tek.Enst.Proje No:DBTE-098.

Anon (1997). Izmir Bay Marine Research 1994–1998 (1997 Report), (in Turkish), D. E. U. Den.Bil.Tek.Enst.Proje No:DBTE-098.

Atılgan, İ. (1997). *Comparative investigations of carbon, organic matter (loss on ignition) and some heavy metal (Cu, Zn) levels of sediments of Gulluk and Homa Lagoon, (in Turkish)* (Yüksek Lisans Tezi, M.Sc.) E. U. Fen Bilimleri Enst. Su Urunleri Anabilim Dalı, pp. 1–15.

Aydın, A. & Sunlu, U. (2005). Investigation of carbon and organic matter (loss on ignition) concentrations in the sediments of South Aegean Sea, (in Turkish), *E.U. Journal of Fisheries and Aquatic Sciences volume*; 21, issue;3–4 pp. 229–234.

Bizsel, N. & Uslu, O., 2000, Phosphate, Nitrogen and Iron Enrichment in The Polluted Izmir Bay, Agean Sea *Mar Environ Res 49 (4): 397-397,*

Büyükışık, B., Ş. Gökpınar & H. Parlak. (1997). Ecological modelling of İzmir Bay. *Journal of Fisheries and Aquatic Sciences, 14(1-2):71-91.*

Colak-Sabancı, F. & T. Koray, 2001, The effect of pollution on the vertical and horizontal distribution of microplankton of the bay of Izmir (Aegean Sea). (in Turkish). *Journal of Fisheries and Aquatic Sciences,* 18(1-2) :187-202.)

Duman, M., Avci, M., Duman, S., Demirkurt, E., & Duzbastilar, M. K. (2004). Surficial sediment distribution and net sediment transport pattern in Izmir Bay, Western Turkey. *Continental Shelf Research,* 24, 965–981.

Egemen, O., Gokpinar, S., Onen, M., Buyukisik, B., Hossucu, B., Sunlu, U., et al. (1999). Gulluk Lagoon ecosystem (Aegean Sea/Turkiye), (in Turkish). *Turkish Journal of Agriculture and Forestry suplement issue,* 3, 927–947.

Environment for Europeans (2005). Keeping our seas alive. *Magazine of the Directorate-General for the environment.* No;21, pp.3–4. Brussels.

Gaudette, H. E., Flight, W. R., Toner, L., & Folger, W. (1974). An inexpensive titration method for the determination of organic carbon in recent sediments. *Journal of Sedimentory Petrology,* 44, 249–253.

Hilal Aydın Gençay & Baha Büyükışık, 2006, The studies on phytoplankton population Dynamics at DEM Harbour (Çandarlı Bay, Aegean Sea), *E.U. Journal of Fisheries & Aquatic Sciences, Volume 23, Issue (1-2): 43–53*

Hitchcock, G. L. (1982). A comparative study of the sizedependent organic composition of marine diatoms and dinoflagellates. *Journal of plankton research,* 4, 363–377

Kaymakci, A., Sunlu, U., & Egemen, O. (2000). Assessment of nutrient pollution caused by land- based activities in Izmir Bay, Turkiye. *Meeting on Interdependency between Agriculture, Urbanization: Conflicts on Sustainable Use of Soil, Water,* Tunisia: pp. 41–49.

Kontas, A., Kucuksezgin, F., Altay, O., & Uluturhan, E., 2004, Monitoring of eutrophication and nutrient limitation in the Izmir Bay (Turkey) before and after Wastewater Treatment Plant. *Environment International,* 29(8), 1057–1062.

Koray, T., Buyukisik, B., Parlak, H. & Gokpınar, Ş.,1992, Unicellular organisms effecting sea water quality in the bay of Izmir: red tides and other blooming, *Doğa Tr. J. of Biology* 16, 135-157 p.

Kucuksezgin, F.; Kontas, A.; Altay, O.; Uluturhan, E. & Darılmaz, E., 2006, Assessment of marine pollution in Izmir Bay: Nütrient, heavy metal and total hydrocarbon concentrations. *Environment International, 32(1), 41-51.*

Kukrer, S., (2009). *Investigating the effects of creeks to eutrophication in İzmir Bay on cleaning process.* Ege University,PhD thesis, 162 p.

Kükrer, S. & Aydın, H. 2006.Investigation of temporal changes of phytoplankton in Karşıyaka Yacht Port (İzmir iner bay). *E.U. Journal of Fisheries & Aquatic Sciences 2006 Volume 23, Issue (1-2): 139-144.* (in Turkish).

Lorenzen, C. J. (1971). Chlorophyll-degradation products in sediments of Black Sea. *WoodsHole Oceanographic Institution Contribution* No; 2828, pp. 426-428.

Morris, I., 1980, *The Physiological Ecology Of Phytoplankton.*(Studies İn Ecology; Vol 7) University of California Press. P. 621

Özkan, E.Y., Kocataş, A. & Büyükışık, B., 2008, Nutrient dynamics between sediment and overlying water in the iner part of Izmir Bay, Eastern Aegean, *Environmental Monitoring* Assess 143, 313-325 p.

Parsons, T. R., Matia, Y., & Lalli, C. M. (1984). *A manual of chemical and biological methods for seawater analysis* p. 173. New York: Pergamon Press.

Scoullos, M., & Dassenakis, M. (1982). Trace metal levels in sea water and sediments of Evoikos Gulf. Greece. *Ives Journe'es Etud. Poll. Cannes, CIESM,* pp. 425–429.

Smayda, T. J. (1997). Harmful algal blooms: Their ecophysiology and general relevance to phytoplankton blooms in the sea. *Limnology and oceanography,* 42(5, part 2), 1137–1153.

Strickland, J. D. H., & Parsons, T. R. (1972). *A practical handbook of seawater analysis.* Fisheries Research Board of Canada. Bull. 167. Ottowa, 310 p.

Sunlu, U., Aydın, A., & Egrihanci (Ozcetin) N. E. (2005). Investigation of carbon and organic matter (%) concentrations in the sediments of North Aegean Sea, (in Turkish), *E.U. Journal of Fisheries & Aquatic Sciences* 22,(3-4), 263-268.

Sunlu, U., Buyukisik, H. B., Koray, T., Brockel, K., Sunlu, F. S. & Sever, T. M. et al. (2007). The effects of Izmir Big Channel waste water treatment project to the lower trophic level of Izmir Bay (Aegean Sea, Turkey). The Scientific and Technical Research Council of Turkey (TUBITAK-CAYDAG) Project No;102Y116 Final Report. 253p. (in Turkish).

Sunlu, U., Egemen, O., Kaymakci, A., & Tüzen (1999). Investigation of impact of fish rearing at net cages on sediment in Urla Pier, (in Turkish), X. National Fisheries Symposium., 22–24 september 1999. Adana.

Turkman, A. (1981). Evaluation of results of the analyses of the secondary rivers discharging to Izmir Bay, (in Turkish), Devolopment of Soil And Water Resources Conference, 2, 723–750.

Varnavas, S. P., & Ferentionos, G. (1982). Heavy Metal distribution in the surface sediments of Pataraikos Bay, Greece, *Vies Journes'es Etud. Poll.Cannes, CIESM.,* pp. 405–409.

Wood, R., (1975). *Hydrobotanical Methods,* Univ.Park Press, *London:173.*

Yaramaz, O., Modogan, H., Onen, M., Sunlu, U., & Alpbaz, A. (1991). Investigation of some heavy metals (Fe,Mn, Ni) and organic matter (%C) concentrations in the sediments of the Izmir Bay, (in Turkish), X. National Fisheries Symposium.12–14 Kasim 1991, Izmir, s: 406–413.

Yaramaz, Ö., Önen, M., Sunlu, U., & Alpbaz., A. (1992). Comparative investigations of some heavy metal (Pb,Cd, Zn,Cu) levels in the sediments of Izmir Bay, (in Turkish), First International environmental protection symposium proceding, vol 2. In Zafer AYVAZ (ed.), s.15–19, Ege University Izmir-TURKIYE.

11

Sewage Sludge Disposal and Applications: Self-heating and Spontaneous Combustion of Compost Piles - Trace Metals Leaching in Volcanic Soils After Sewage Sludge Disposal

Mauricio Escudey[1,4], Nelson Moraga[2],
Carlos Zambra[2] and Mónica Antilén[3,4]
[1]Universidad de Santiago de Chile, Facultad de Química y Biología, Departamento de Química de los Materiales, Av. B. O'Higgins 3363, Santiago,
[2]Universidad de Santiago de Chile, Facultad de Ingeniería, Departamento de Ingeniería Mecánica, Av. Bdo. O'Higgins 3330,
[3]Pontificia Universidad Católica de Chile, Facultad de Química, Departamento de Química Inorgánica, Vicuña Mackenna 4860, Santiago,
[4]Center for the Development of Nanoscience and Nanotechnology, CEDENNA, Santiago, Chile

1. Introduction

Municipal solid waste landfills often develop scenarios of self heating, causing negative environmental impacts by odors, gas generation and smoke production. Self-ignition and resulting fires at landfills have been undesirable outcomes in compost piles worldwide. Field, laboratory and numerical studies have been considered trying to reproduce and understand the conditions under which self-heating and combustion may take place. Inside a compost pile built from solids obtained after municipal wastewater treatment, oxygen, methane, carbon monoxide and carbon dioxide concentrations and temperature change with time and in depth. Electric conductivity and pH show only slight changes. In field piles temperature increased with time, reaching a maximum of about 90°C. While no spontaneous combustion was observed after six months in field experiments, in laboratory studies carried out in a closed bottom cylinder, self-ignition was observed and a maximum temperature of about 400°C was reached. Spontaneous combustion in compost piles is a chemical and biological process. Initially, the metabolism that allows microorganism growth causes temperature increase, but simultaneous oxidation of the organic matter also reinforces self-ignition at a lower value. When the temperature within the compost pile rises to about 87°C, self-ignition follows mainly due to cellulose oxidation. Therefore, the conditions under which biomass increases must be studied. An important factor in the development of internal changes within the compost pile comes from atmospheric boundary conditions. General convection, radiation, rain rate and atmospheric humidity must be included as boundary atmospheric field conditions on the lateral and top surfaces.

Local thermal equilibrium is assumed, which is a common assumption for porous media and packed particle beds, hence effective properties can describe the porous media. Coupled two-dimensional heat and oxygen diffusion in a compost pile of sewage sludge obtained from domestic wastewater treatment can be described using a transient, nonlinear mathematical model that includes the volumetric heat generation caused by the action of aerobic bacteria and by the oxidation of cellulose in a porous medium. Numerical simulations with the finite volume method allow the prediction of the pile's shape and size effects on the heat generated and on oxygen consumption. Transient temperature and oxygen concentration distributions within compost piles depend on their geometry. Heat (temperature) and mass transfer (oxygen) results indicate that pile height has an important effect on the heating. The volumetric heat generation caused by the action of aerobic bacteria and by the oxidation of cellulose in a porous medium must be incorporated as a function of the mean moisture by an essentially thermodynamic source term. Inside a compost pile made of solid residues obtained after municipal wastewater treatment, temperature changes, oxygen content and moisture can be measured in time and depth. The Richards equation is a standard, frequently used approach for modeling and describing water flow in variably saturated porous media such as soils and compost piles. The nonlinear mathematical model considers the Richards equation and a second mass transport equation for water content, and therefore the numerical simulations can describe the internal changes in heat, oxygen and moisture observed under field conditions. Therefore a model with three diffusion equations must be used to quantify the effects of moisture on temperature and oxygen concentration.

1.2 Materials and methods. Experimental conditions

For field and laboratory studies sewage sludge from "El Trebal", a domestic wastewater treatment plant located close to the city of Santiago, Chile, was used, where about 4 m^3/s of wastewater are treated. In the field, four sewage sludge piles 1.5 and 2.5 m high, 6.2-7 m wide, and 8.5 m long, with a 3-D trapezoidal shape and a bulk density of 0.75 ton/m^3 were prepared. Temperatures were measured using type K thermocouples (Ni-Cr) connected to an LA/AI-A8 Cole Parmer data acquisition system, and they were recorded every 5 min, during 80 to 120 hours. Measurements of humidity, O_2, CH_4, CO_2, and CO were made twice a week; pH, electric conductivity, and organic carbon were determined once a week. Water content was determined from the weight lost after heating 10 g of sample at 105 °C for 24 hours, and electrical conductivity and pH were determined in a 1:2.5 solid:water ratio extract. Organic carbon content (expressed as %OC), was determined by a wet oxidation method (Walkley & Black, 1934) where the sample reacts with a mixture of 1 N $K_2Cr_2O_7$ and concentrated H_2SO_4.

A Dräger Miniwarn equipment, equipped with an IR catalytic sensor for CH_4 and electrochemical sensors for O_2, CO_2 and CO, was used considering a 1.5 L/min mass flow. Field experiments were carried out for 20 weeks, a period during which wind speed and direction, relative humidity of air, and precipitation were also recorded.

In the laboratory a stainless steel cylinder 20 cm in diameter, 20 cm deep, and 2 mm wall thickness was used. The cylinder was externally insulated, with openings at depths of 0.0, 1.0, 2.0, 2.5, 3.0, 7.0, 10.0 and 15.0 cm for thermocouples location. An adjustable electric heater was incorporated 2.5 cm over the cylinder, and experiments were carried out at different external temperatures. Sewage sludge was placed in the cylinder at the same bulk density observed in the field.

1.3 Mathematical modeling and numerical simulation of compost pile. Self-heating in compost pile, differential equations for temperature and oxygen concentration.

Temperature and oxygen concentration are the important dependent variables for describing heat and mass transfer processes inside compost piles. Basic equations to build appropriate mathematical models, numerical solutions procedures, and the results obtained are presented in this section.

The mathematical model considers transient coupled two-dimensional heat and oxygen diffusion in porous media (Sidhu et al., 2007). The pile bottom is assumed to be adiabatic, as shown in Figure 1.

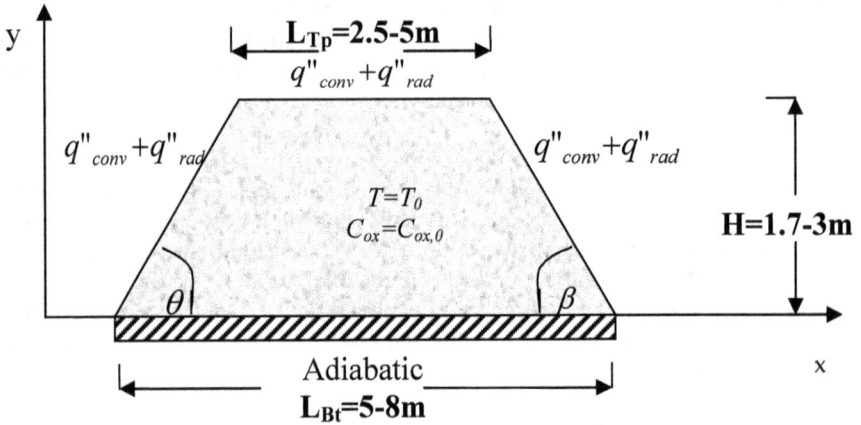

Fig. 1. Physical situation of the compost pile

Cellulosic oxidation and micro organism activity inside the pile are incorporated in the model by volumetric heat generation. For simplicity, local thermal equilibrium is assumed, which is a common assumption for porous medium and packed particle beds (Nield & Bejan, 1992).

$$\frac{\partial(\rho C)_{eff} T}{\partial t} = \nabla k_{eff} \nabla T + Q_c(1-\varepsilon)A_c\rho_c C_{ox} \exp(\frac{-E_c}{RT}) + Q_b(1-\varepsilon)\rho_b\rho_c(\frac{A_1 \exp(\frac{-E_1}{RT})}{1+A_2 \exp(\frac{-E_2}{RT})}) \qquad (1)$$

$$\varepsilon\frac{\partial C_{ox}}{\partial t} = \nabla D_{eff}\nabla C_{ox} - (1-\varepsilon)A_c\rho_c C_{ox} \exp\left[\frac{-E_c}{RT}\right] \qquad (2)$$

In equations (1) and (2) A_c is the pre-exponential factor for the oxidation of the cellulose and E_c, E_1, E_2, are the activation energy for cellulose, biomass growth and inhibition of biomass growth, respectively. Heat and mass transfer properties in the porous media are defined in terms of the pile's porosity ε,

$$k_{eff} = \varepsilon k_{air} + (1-\varepsilon)k_c \qquad (3)$$

$$(\rho C)_{eff} = \varepsilon\rho_{air}C_{air} + (1-\varepsilon)\rho_c C_c \qquad (4)$$

$$D_{eff} = \varepsilon D_{air,c} \tag{5}$$

where k_{eff} and D_{eff} are the effective properties which are considered independent of temperature and concentration, and C_c is the specific heat capacity of cellulose. The heat produced by the oxidation of the cellulosic material is represented by the second term on the right-hand side of the Equation (1). This term is expressed as a function of oxygen concentration and is responsible for the self-heating. The last term in Equation (1) represents heat generated due to biological activity within the pile, caused by microorganism growth. Oxygen content variation is included in Equation (2). Initially, oxygen content in the pile has the same concentration as in the external air. Micro-organisms growth followed by organic matter oxidation produces the oxygen depletion. These assumptions are incorporated in the second term on the right-hand side of the Equation (2).

Details in the formulation of the term representing the heat generated by the biomass have been given by Chen and Mitchell (Chen & Mitchell, 1996) and some parameter values used in the mathematical model may be obtained from the literature (Sidhu et al., 2007; Escudey et al., 2008; Moraga et al., 2009).

1.4 Temperature and oxygen concentration, initial and boundary conditions
Initial temperature and oxygen and moisture distribution within the pile are assumed starting from the corresponding first experimental values available:

$$T(x,y,0) = T_0 ; \qquad C_{ox}(x,y,0) = C_{ox,0} \tag{6}$$

Adiabatic and impermeable boundary conditions are considered at the pile's base

$$y = 0: \qquad \left.\frac{\partial T}{\partial y}\right|_{y=0} = 0 ; \qquad \left.\frac{\partial C_{ox}}{\partial y}\right|_{y=0} = 0 \tag{7}$$

Exchange with the surroundings at the outer pile surface should consider heat transfer caused by a combined action due to convection and radiation through impermeable walls:

$$y = H: \qquad -k_{eff}\left.\frac{\partial T}{\partial y}\right|_{y=H} = q''_{conv} + q''_{rad} ; \quad D_{eff}\left.\frac{\partial C_{ox}}{\partial y}\right|_{y=H} = 0 \tag{8}$$

$$-k_{eff}\left.\frac{\partial T}{\partial n}\right|_{y=H} = q''_{conv} + q''_{rad} ; \quad D_{eff}\left.\frac{\partial C_{ox}}{\partial n}\right|_{y=H} = 0 \tag{9}$$

$$q''_{conv} = h \cdot \left[T - T_a(t)\right] \tag{10}$$

The heat transfer convective coefficient can be assumed to change with ambient air velocity v (Duffie & Beckman, 1980):

$$h = 5.7 + 3.8 * v \tag{11}$$

External thermal radiation incorporating incoming daily solar radiation and nocturnal heat losses can be calculated in terms of a function that periodically changes with time according to

$$q''_{rad} = a \cdot sen(w \cdot t) \tag{12}$$

The pile dimensions indicate that an idealized two-dimensional domain is appropriate to assess the effects of pile geometry. Three cases were considered for the compost piles investigated: 1) symmetric trapezoidal; 2) asymmetric trapezoidal; 3) fifth polynomial contour.

1.5 Numerical prediction using the Finite Volume Method.

The system of equations that governs this problem (1-12) can be solved numerically using the finite volume method, Patankar (Patankar, 1991). Each one of the governing equations was written in the general form of the transport equation, with unsteady, diffusion and linearized source terms:

$$\frac{\partial \phi}{\partial t} = div(\Gamma \cdot grad\phi) + Sc + Sp \cdot \phi \tag{13}$$

First-order accuracy in time was used in the numerical scheme to account for the unsteady heat and mass terms:

$$\frac{\partial \phi}{\partial t} = \frac{\phi^{t+\Delta t} - \phi^t}{\Delta t} \tag{14}$$

The diffusion coefficient (Γ) and source terms (Sc, Sp) for each dependent variable ϕ are given in Table 1.

ϕ	Γ	S_C	S_P
T	k_{eff}	$\dfrac{T^p}{\Delta t} + Q_c(1-\varepsilon)\rho_c A_c C_{ox} \exp(\dfrac{-E_C}{RT}) +$ $Q_b(1-\varepsilon)\rho_b\rho_c(\dfrac{A_1 \exp(\dfrac{-E_1}{RT})}{1 + A_2 \exp(\dfrac{-E_2}{RT})})$	$-\dfrac{1}{\Delta t}$
C_{ox}	D_{eff}	$\dfrac{C_{ox}^p}{\Delta t}$	$-(1-\varepsilon)A_c\rho_c \exp\left[\dfrac{-E_c}{RT}\right] - \dfrac{1}{\Delta t}$

Table 1. Diffusion coefficient and source terms for the mathematical model.

An original computer program written in Fortran, with a combination of the TDMA and the iterative Gauss-Seidel method (Patankar, 1991), was used to predict temperature and oxygen concentration inside the compost pile.

The pile geometry was discretized using three uniform grids: 100x100, 200x200 and 300x300 nodes in the x and y directions, respectively, to verify that the results obtained were not influenced by mesh size. Numerical simulations were carried out for four piles: 1.7, 1.8, 2.5 and 3.0 m height.

Temperature was noticed to increase initially smoothly with time while oxygen decreased slowly with time until sudden changes, caused by the cellulosic heat generation, were noticed for both temperature and oxygen concentration.

Therefore, a strategy based on the use of dynamic time steps was implemented, with lower time steps when the unsteady terms were higher. The different values of the time steps used in the unsteady calculations were in the interval

$$300s \leq \Delta t \leq 3600s \tag{15}$$

The iterative procedure ended at each time step when the maximum difference between iteration for $\phi = T, C_{ox}$ satisfied, at each control volume, the convergence criteria

$$\phi_{i,j}^{k+1} - \phi_{i,j}^{k} \leq 10^{-4} \tag{16}$$

1.6 Comparison between numerical predictions and experimental results.

In order to validate the mathematical model and the numerical simulations a 2.5 m high, 8.5 m long and 7.0 m wide compost pile was built, with a 3D trapezoidal shape, with a 2.5 m wide top surface. Temperature was recorded using type K thermocouples (Ni-Cr) 0.1 m long and 0.0015 m in diameter, built with a type G junction and a silver end, provided with a 5-m long cable with flexible stainless steel coating. The thermocouples were connected to an LA/AI-A8 Cole Parmer data acquisition system coupled with a PC computer provided with the InstaTrend software programmed to collect temperatures every 10 minutes. Every week the values of the computer collected data were averaged and plotted. The temperature measurement points were located in the pile center (z axis) at 2.1, 1.25 and 0.35 m depth. The pile was built with sewage sludge produced by a municipal wastewater treatment plant located in the city of Santiago, Chile (El Trebal) in a landfill belonging to the same plant. The sewage sludge was produced in July 2004 and was used to build the pile in the third week of February 2005.

The experimental and numerical data are compared in Figure 2 (Escudey et al., 2008). Data obtained from numerical calculations were plotted considering daily output at 12:00 a.m. The experimental and predicted data follow the same general trends. Near the surface (0.35 m depth) during the third week the main differences found were not greater than 3°C. The best description of the experimental data was obtained at a depth of 2.1 m.

1.7 General analysis results for the temperature time evolution in a compost pile.

A fundamental process in a compost pile is to achieve a higher enough temperature to cause the death of pathogenic micro-organisms together with the degradation of carbonaceous material. Field practice indicates that an adequate compost process may be achieved when the temperature inside the pile is between 313 K and 353 K. Initially, the temperature inside the pile increases slowly from 293 K up to 353 K due to microorganism growth, as described by the last term on the right-hand side of Equation (1). As the temperature increases beyond 353 K the death of different microorganism colonies is followed by a sudden increase in cellulose oxidation, included as a volumetric heat generation in the second term on the right-hand side of Equation (1). Cellulose oxidation is accounted for in terms of the oxygen concentrations in the heat generation term of Equation (1) and of its counterpart term of oxygen depletion in the last term of the mass diffusion Equation (2). The sudden temperature increase is known as a thermal explosion.

The compost's physical properties change dramatically when the soil has undergone a thermal explosion and hence the compost material cannot be used as a soil fertilizer. Field control of such a sudden temperature increase is a very difficult task and thermal explosions often occur in practice, with loss of the compost material.

A first case of self-heating in a rectangular porous pile 2.5 m high (H) and 5 m long (L) was investigated using three grids, with 100x100, 200x200 and 300x300 nodes, and three time steps: 300 s, 600 s and 3600 s. The temperature-time evolution was calculated at the three positions: f) H/4, L/4; g) H/2, L/2, and h) 3H/4, 3L/4. The results of the time in days needed to cause self-ignition are shown in Table 2 for the three positions. The use of a time step of 600 s and a grid with 300x300 nodes allows to calculate self ignition times of 247, 246 and 250 days at the three vertical positions, respectively, regardless of the time and space discretizations.

Grids	Δt (s)	f) H/4,L/4	g) H/2,L/2	h) 3H/4, 3L/4
100x100	3600	247	245	249
	600	248	248	250
	300	244	241	246
200x200	3600	248	246	251
	600	250	249	253
	300	246	244	249
300x300	3600	248	247	252
	600	247	246	250
	300	247	246	250

Table 2. Days before self-ignition at positions f), g) and h) within the compost pile.

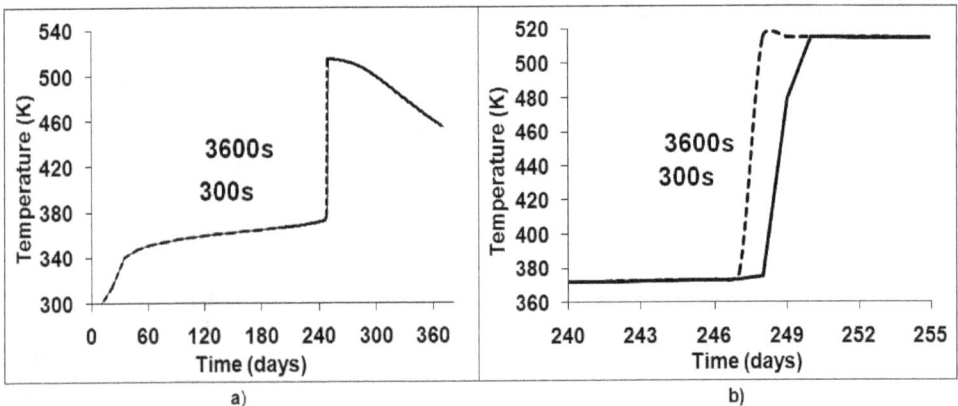

Fig. 3. Temperature evolution in time calculated with two time steps, 300x300 mesh, at position H/4, L/4, for a 2.5m high pile, a) full time scale, b) during thermal explosion.

Figure 3 shows the time evolution results of temperature at the f) position: H/4, L/4, calculated with a grid of 300x300 nodes using two time steps: 300 s and 3600 s. A typical heating curve depicting four stages is observed in Figure 3a. A first phase, lasting 35 days, where temperature increases from 293 K to 340 K, is seen. The growth and activities caused by aerobic micro-organisms, included in the mathematical model by a volumetric heat generation, originated the initial pile heating. A second phase, from day 35 to day 246, with a very slow heating process in which temperature increases from 340 K to 370 K, is originated initially by thermophilic micro-organisms whose number decreases progressively as temperature increases, followed by the cellulosic oxidation of the wood chips (used to increase the pile's mechanical strength), incorporated as the first heat generation term in the right-hand side of Eq. (1). Auto-ignition at H/4, L/4 occurs after 247 days, during the third stage, in which temperature increased suddenly in one day from 370 to 515 K. A complex system of solid, liquid and gaseous fuels as a final result of cellulosic oxidation caused a volumetric heat generation that led to the self-ignition process. Temperature decreasing in time characterized the fourth and last stage, in which the fuel reserves at the location are becoming exhausted.

Figure 3b, a zoom view of Figure 3a for the time interval between days 240 and 255, shows that a time step reduction from 3600 s to 300 s allows a more accurate prediction (within 1 day) of the time needed to initiate the self-ignition in a 2.5 m high compost pile.

From the above analysis, a mesh with 300x300 nodes and a dynamic time step of 300 s during the self-ignition and 3600 s in the other stage are needed to solve the problem in an efficient way.

1.8 Pile height effect on temperature and oxygen concentration

In this section a critical pile height of 1.8 m for the thermal explosion to occur has been found. Heating curves at three different positions (H/4, H/2 and 3H/4) located in the mid horizontal section of four rectangular piles (5 m wide) with 1.7, 1.8, 2.5 and 3.0 m heights are shown in Figure 4. Temperature is observed to increase to 369 K in the 1.7 m high pile, at the H/4 and H/2 vertical locations, and to 353 K at the 3H/4 position after 728 days. A sudden temperature increase that took place at around 240 days for the 2.5 and 3.0 m high piles, and at 397 days (in H/4), 374 days (in H/2) ,and 409 days (in 3H/4) can be seen to characterize the initial time for the self-ignition process. Compost piles with heights greater than 1.7 m were found to initiate self-ignition in a time that is inversely proportional to pile height.

Changes in oxygen content with time, calculated inside the 1.7, 1.8, 2.5 and 3.0 m high compost piles, are described in Figure 5 at three vertical positions (H/4, H/2 and 3H/4). As a result of the coupling between heat and mass transfer caused by the cellulose oxidation in both diffusion equations, sudden oxygen depletions are observed at the 1.8, 2.5 and 3.0 m pile height positions, occurring at the time when fast temperature increases were noticed. When pile height is 1.7 m, oxygen concentration inside the pile remains almost constant during the 728 day period used in the analysis. Oxygen reduction to low levels in the domain, coupled with the high temperatures achieved in the same regions, destroys the aerobic thermophilic micro-organisms.

Table 3 summarizes the results obtained after the self-ignition analysis calculated from the time evolution of temperature and the oxygen concentration distributions inside the four rectangular piles of different heights. Temperature and oxygen distribution results show that when the pile volume to heat transfer external area ratio, V/A, is greater than or equal to 1.05 self-ignition occurs inside the pile.

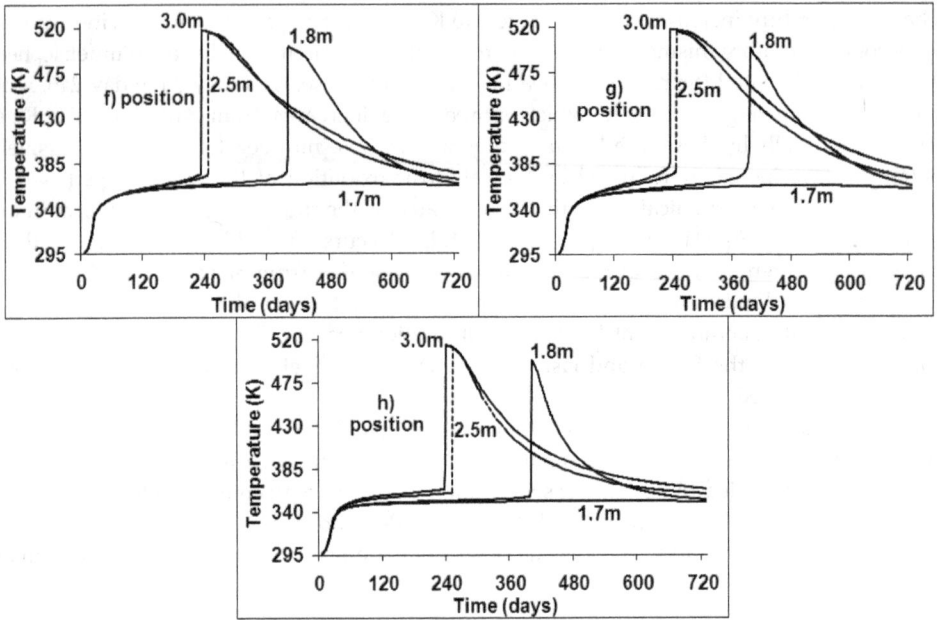

Fig. 4. Time evolution of temperature for four piles with different heights.

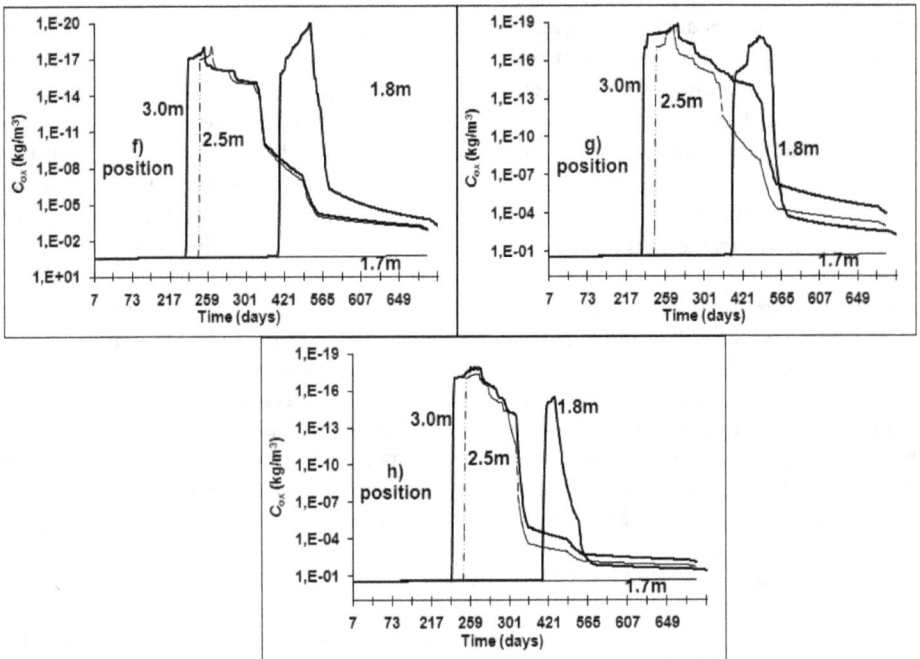

Fig. 5. Time evolution of oxygen concentrations for four piles with different heights.

Pile Height (m)	Area (m²)	Volume (m³)	V/A Ratio (m)	Self-Ignition
1.7	8.4	8.5	1.01	no occur
1.8	8.6	9	1.05	occur
2.5	10	12.5	1.25	occur
3.0	11	15	1.36	occur

Table 3. V/A ratio in four piles of different height.

Micro-organism and cellulosic heat generations inside the piles, calculated at a position located at mid section and 3H/4 for four different heights are presented in Figure 6 for the rectangular pile. A rapid initial growth followed by a fast fall in the microorganism volumetric heat generated can be noticed for the four pile heights studied. While cellulosic heat generation inside a 1.7 m pile height is seen to increase to a value that remains almost constant after 450 days, piles taller or equal to 1.8 m show a monotonous increment that is followed by a sudden increment, causing self-ignition. The heat generated by the biological activity of micro-organisms becomes equal to the heat generated by the cellulose oxidation after 102 days for the 1.7 m pile, 105 days for the 1.8 m, 84 days for the 2.5 m, and 77 days for the 3.0 m high piles. In the 1.7 m high pile, after 100 days, cellulosic oxidation keeps the temperatures for all time simulation while the oxygen consumption increases until day 500 and then decreases. Maximum heat generation values for the cellulose oxidations are 2.4×10^5 (W/m³), 5.7×10^5 (W/m³) and 6.4×10^5 (W/m³) for the 1.8 m, 2.5 m and 3.0 m high piles, respectively.

In the water treatment industry the piles are about 1.5 m high and 120 m long. Numerical results verified by experimental observations show that temperature increase in the pile is a function of the volume. Therefore, pile height must be considered a very important factor in thermal explosions. When the initial oxygen content within the pile is at adequate concentrations, microorganism and temperature increases are seen. This occurs until the temperature reaches 353 K, when two phenomena occur: death of the micro-organisms and cellulosic oxidation. The numerical results shown in Figures 6-8 reproduce the biochemical processes described above, resulting in sudden changes in both oxygen and temperature in piles 1.8 m high or higher.

1.9 Compost pile geometric effects.

Heat and mass transfer characteristics are described in Figures 9-11 in terms of temperature and oxygen concentration distribution at four different times in the neighborhood of the self-ignition time. These results were obtained by numerical simulations using the proposed model, equations (1-12). Figure 9 shows the temperature-time evolution and oxygen concentration distributions for a symmetrical trapezoidal pile 3 m high and 8 m at the base, with lateral walls inclined at 45°. A maximum temperature of 440 K was reached after 208 days in the pile's lower-central section, which increases to 519 K on day 216 and is located in the central region at H/3. Temperatures higher than 500 K were measured in the central pile region for 60 days. Oxygen content inside the pile is observed to tend to zero in the lower central section of the pile from 216 to 252 days. Almost the whole pile was seen to be undergoing self-ignition on day 225 except in the regions close to the lateral and top areas.

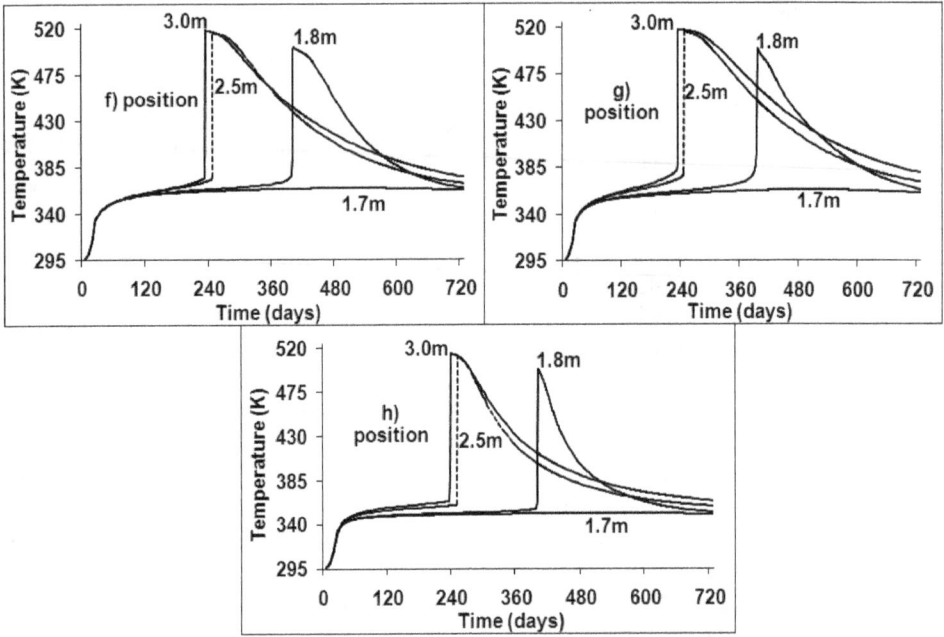

Fig. 6. Time evolution of temperature for four piles with different heights.

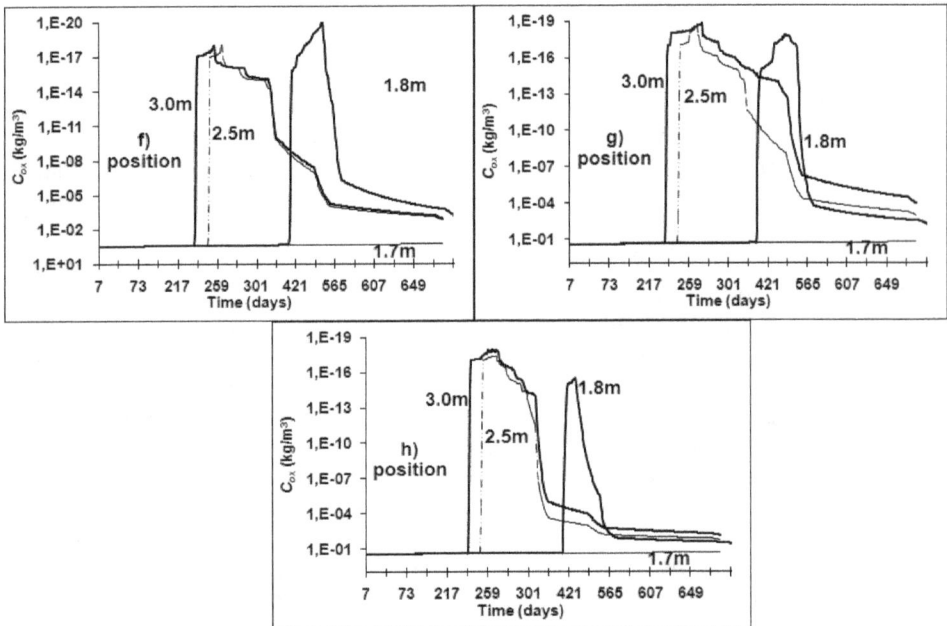

Fig 7. Time evolution of oxygen concentration for four piles with different heights.

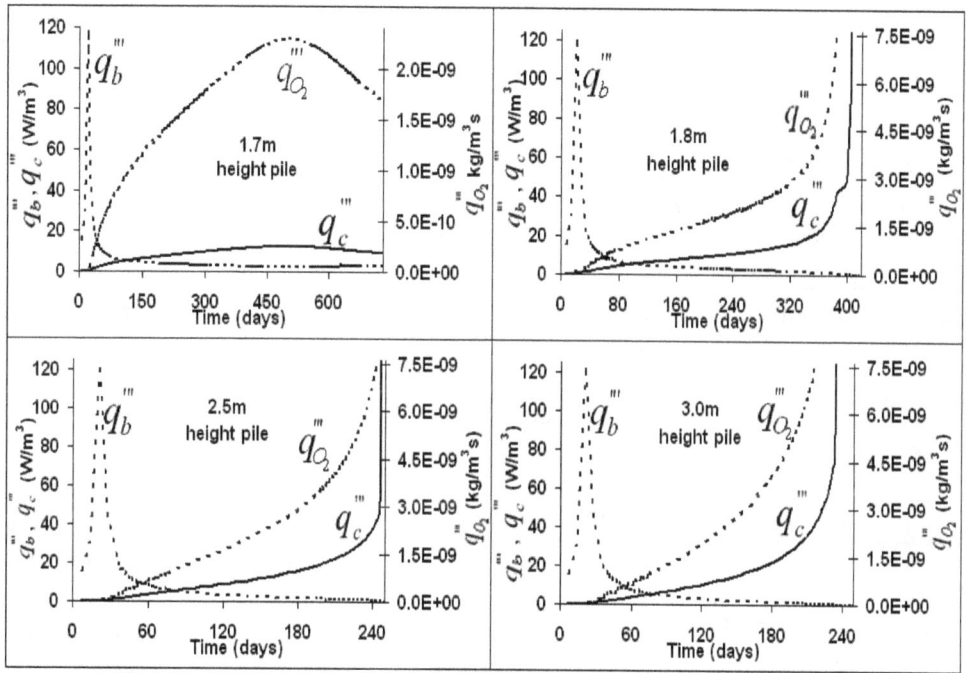

Fig. 8. Time evolution of energy and oxygen consumption for four piles with different heights.

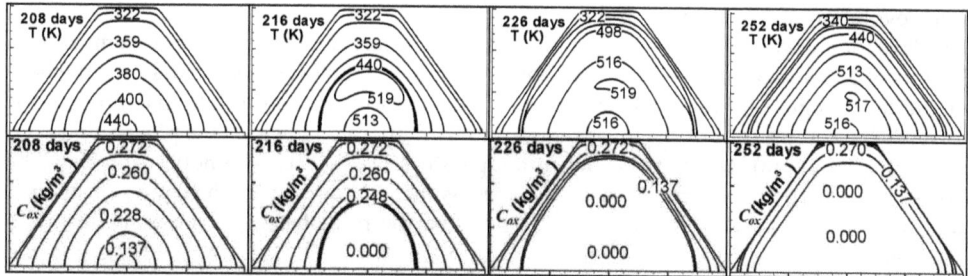

Fig. 9. Distribution of temperature and oxygen concentration within a compost pile, trapezoidal geometry.

Heat and mass transfer inside a non-symmetric trapezoidal sludge pile with an 8x8 m base and lateral walls inclined at angles of $\theta = 56.3°$ and $\beta = 33.7°$ caused by chemical and biological reactions, are described in Figure 10 in terms of temperature and oxygen distribution. Self-ignition started on day 215, when a maximum temperature of 513 K was achieved. A narrow region with high temperature gradients can be observed in the lower central region of the pile on day 217. At this time the self-ignition front is closer to the bigger lateral wall (with $\beta = 33.7°$), and therefore smoke production can be expected to begin there. During self ignition the zone with maximum temperature, between 516 and 519 K, reached between days 217 and 253 is closer to the shorter lateral wall.

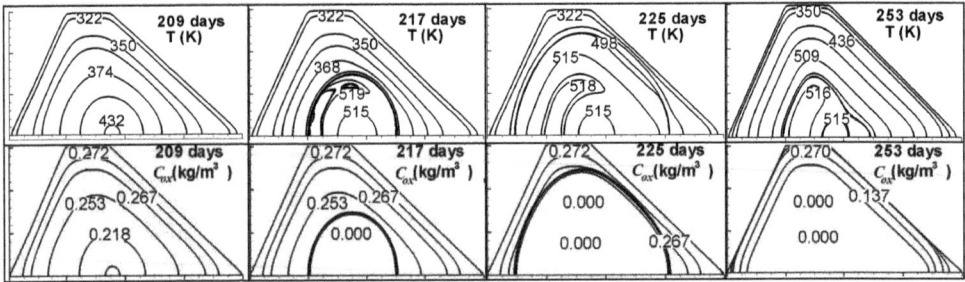

Fig. 10. Distribution of temperature and oxygen concentration within a compost pile, asymmetric geometry.

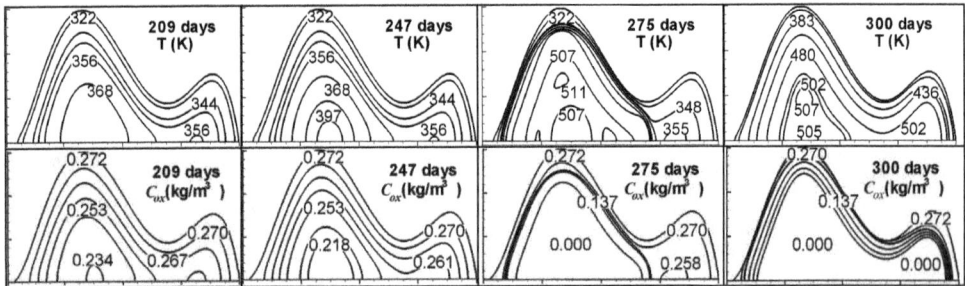

Fig. 11. Distribution of temperature and oxygen concentration within a compost pile, polynomial geometry.

Time evolution of temperature and oxygen concentration distributions at 209, 247, 275 and 300 days for an asymmetrical compost pile with two different height bumps, with maximum heights of 3 m and 8 m at the base, are shown in Figure 11. Self-ignition occurs near the base of the taller region on day 258, with a maximum temperature of 493 K that propagated towards the central zone of the taller region and then migrated towards the pile section with lower height (1.5 m), where a maximum temperature of 502 K can be noticed on day 300. The self-combustion zone can be easily detected as the region in which the oxygen content is zero and on day 300 it can be seen to extend from the pile's base to a region close to the external walls.

1.10 Flow in compost pile as an unsaturated porous medium

The Richards equation (RE) (Richards, 1931) is a standard, frequently used approach for modeling and describing flow in variably saturated porous media. RE is obtained by combining Darcy–Buckingham's law with the mass conservation or continuity equation, under the assumption that the air phase remains at constant (atmospheric) pressure and the water phase is incompressible. Using one dimensional flow in a vertical direction, y, as an example, the following equations depict Darcy's and continuity equation, respectively.

$$q_D = -k_h \frac{\partial \Psi}{\partial y} = -k_h \frac{\partial(\psi + y)}{\partial y} = -k_h \left(\frac{\partial \psi}{\partial y} + 1 \right) \quad (17)$$

$$\frac{\partial \theta}{\partial t} = -\frac{\partial q_D}{\partial y} \pm S \tag{18}$$

where q_D is the flux density (m/s), k_h is the hydraulic conductivity (m), Ψ is the head equivalent of hydraulic potential (m), is the head pressure (m), θ is the volumetric water content (m³/m³), y is the vertical coordinate, t is time (s) and S is the source term. Substitution of Equation (17) into (18) gives the mixed formulation of RE:

$$\frac{\partial \theta}{\partial t} = \frac{\partial}{\partial y}\left[k_h \left(\frac{\partial \psi}{\partial y} + 1 \right) \right] \pm S \tag{19}$$

Introducing a new term, $D(\theta)$ into (19) gives the soil moisture based form of RE. $D(\theta)$ is the ratio of the hydraulic conductivity, and the differential water capacity is therefore defined as

$$D_{(\theta)} = \frac{k_h}{\frac{\partial \theta}{\partial \psi}} = k_h \frac{\partial \psi}{\partial \theta} \tag{20}$$

$D(\theta)$ is a function of moisture content. This dependence is obtained from field tests. Combining Equations (19) and (20) gives the θ − based form of RE:

$$\frac{\partial \theta}{\partial t} = \frac{\partial}{\partial y}\left(D_{(\theta)} \frac{\partial \theta}{\partial y} \right) + \frac{\partial k_h}{\partial y} \pm S \tag{21}$$

If the gravitational and the source term effects are neglected, the $\frac{\partial k_h}{\partial y}$ the and S terms in Equation (21) are equal to zero.

$$\frac{\partial \theta}{\partial t} = \nabla D_{(\theta)} \nabla \theta, \tag{22}$$

The volumetric water content is the quotient between water volume and total sample volume, so it has no units and its values are between 0 and 1.

The 1D mass transfer of water in soil solution of Equation (22) for volumetric water content diffusion, testing the effects on the thermal properties caused by moisture in porous media, has been reported by Serrano (Serrano, 2004). This diffusivity coefficient of water in a compost pile is calculated by a nonlinear equation:

$$D_{(\theta)} = \vartheta_1 e^{\lambda(\theta)^\alpha} - \vartheta_2, \tag{23}$$

The constants ϑ_1, ϑ_2, λ and α can be obtained by experimental field tests (Serrano, 2004). Equations (22) and (23) may be used when the specific hydraulic properties of the compost pile are not available.

The effects of the vaporization of water on the internal energy may be calculated by incorporating the third term of the right hand side of Equation (17):

$$-L_v \rho_{va} q(\theta) X_v \tag{24}$$

where L_v is the vaporization enthalpy, $\rho_{v,a}$ is the water vapor density, $q(\theta)$ is the mass water flux, and X_v is the vapor quality.

1.11 Humidity, initial and boundary conditions.
Moisture distributions within the pile are assumed starting from the corresponding first experimental values available:

$$\theta(x, y, 0) = \theta_0 \tag{25}$$

A constant volumetric concentration was imposed at the pile base:

$$\left.\frac{\partial \theta}{\partial y}\right|_{y=0} = 0.6. \tag{26}$$

Heat transfer to the environment when the liquid – vapor phase change takes place was calculated with the equation

$$q''_w = h_w L_v (\theta_{w,ml} - \theta_{w,air,ex}) \rho_{w,va}, \tag{27}$$

where $\theta_{w,ml}$ is the water content in the fluid adjacent to the surface, $\theta_{w,air,ex}$ is the water content in the outside air, $\rho_{w,va}$ is the water vapor density on the surface, and h_w is the convective mass transfer coefficient. In order to improve the accuracy of the approximation, q''_w were written in the form of a three-point formula (Ozisik, 1994). On the pile's surface h_w and $\theta_{w,air,ex}$ are affected by the distribution coefficient, K, at the interface between the fluid and the solid. Figure 12 shows three concentration points at the interface used for calculating the mass transfer and convective mass transfer at the solid surface using the equilibrium distribution coefficient (Geankoplis, 1993).

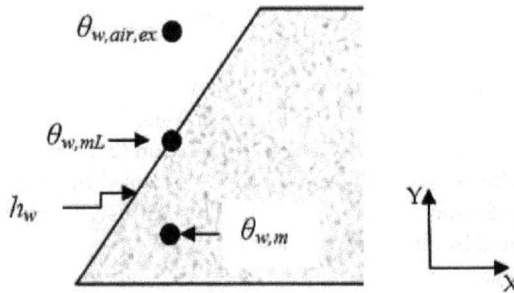

Fig. 12. Source term values for thermodynamic equilibrium

$$K = \frac{\theta_{w,ml}}{\theta_{w,m}}. \tag{28}$$

In Equation (28) $\theta_{w,m}$ is the water content in the solid adjacent to the surface. Substituting Equation (28) in Equation (27) we get:

$$q''_w = h_w K L_v \left(\theta_{w,m} - \frac{\theta_{w,air,ex}}{K} \right) \rho_{w,va} \tag{29}$$

The equation for h_w values is obtained as follows (Kaya et al., 2006):

$$h_w = h\left(\frac{D_{w,air}Le^g}{k_{eff}}\right); \qquad g = 1/3; \qquad Le = k/(\rho \cdot C_p) \cdot D_{w,air} \; (Lewis\; number) \quad (30)$$

Water content at the air-compost interface is calculated assuming an ideal gas mixture and molar concentration depending on partial vapor pressure and temperature at the interface

$$\theta_{w,ml} = \frac{p_{va} * 18}{RT\rho_{w,va}}. \tag{31}$$

The vapor pressure is obtained from the relative humidity, %H, as follows:

$$p_{va} = \%H \cdot p_{va}^*, \tag{32}$$

where p_{va}^* is the saturated vapor pressure. Rain effects as boundary condition were incorporated through Equation (20), considering a relative humidity equal to 1 at the surface of the compost pile.

Convective boundary conditions for water content are introduced through equations (33) and (34).

$$D_{w,air}\left.\frac{\partial\theta}{\partial y}\right|_{y=H} = h_w K\left(\theta_{w,m} - \frac{\theta_{w,air,ext}}{K}\right) \tag{33}$$

$$D_{w,air}\left.\frac{\partial\theta}{\partial y}\right|_{y=H} = h_w K\left(\theta_{w,m} - \frac{\theta_{w,air,ext}}{K}\right) \tag{34}$$

1.12 Experimental and numerical results for humidity.

Unsteady water diffusion inside the sewage sludge was investigated by physical experiments and finite volume simulations, based on the mathematical model described by equations (1-5). A compost pile 2.5 m high, 8.5 m long and 7.0 m wide was built, with a 3D trapezoidal shape and a 2.5 m wide top surface.

Figure 13 shows the values measured for rain, wind velocity, and relative humidity at El Trebal. In the southern hemisphere February–March correspond to the summer season and April–June correspond to the fall season, where ambient temperature decreases from 293 to 282 K. In this period wind speed drops from about 4.5 to 2 m/s and the relative humidity of the air increases from 54 to 84 percent. The frequency and amount of rain also increase in this period, with maximum values of 40 mm in one continuous rainfall period.

The values measured in the field and those calculated by the FVM (Finite Volume Method) for water content, oxygen concentration, and temperature for each point of measurement are shown in Figure 12. Water content from 0.45 to 0.6 represents optimal conditions for biomass growth. In the field experiment those limiting values were exceeded.

The water content in Figure 14.a is affected by the atmospheric conditions of relative humidity and precipitation, and this is clearly seen at a height of 2 m. Further increases in water content within the compost pile take place when both the relative humidity and precipitation (frequency and quantity) increase. Water content at 0.5 m is less affected by the

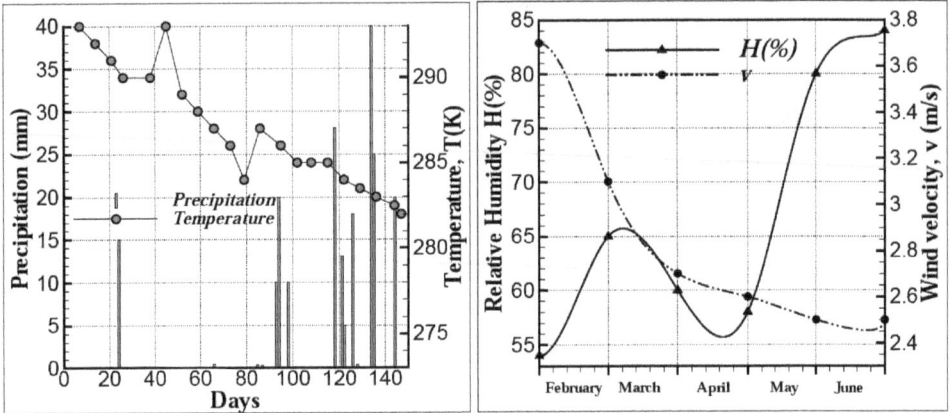

Fig. 13. Precipitation, temperature, relative humidity and wind in *El Trebal*.

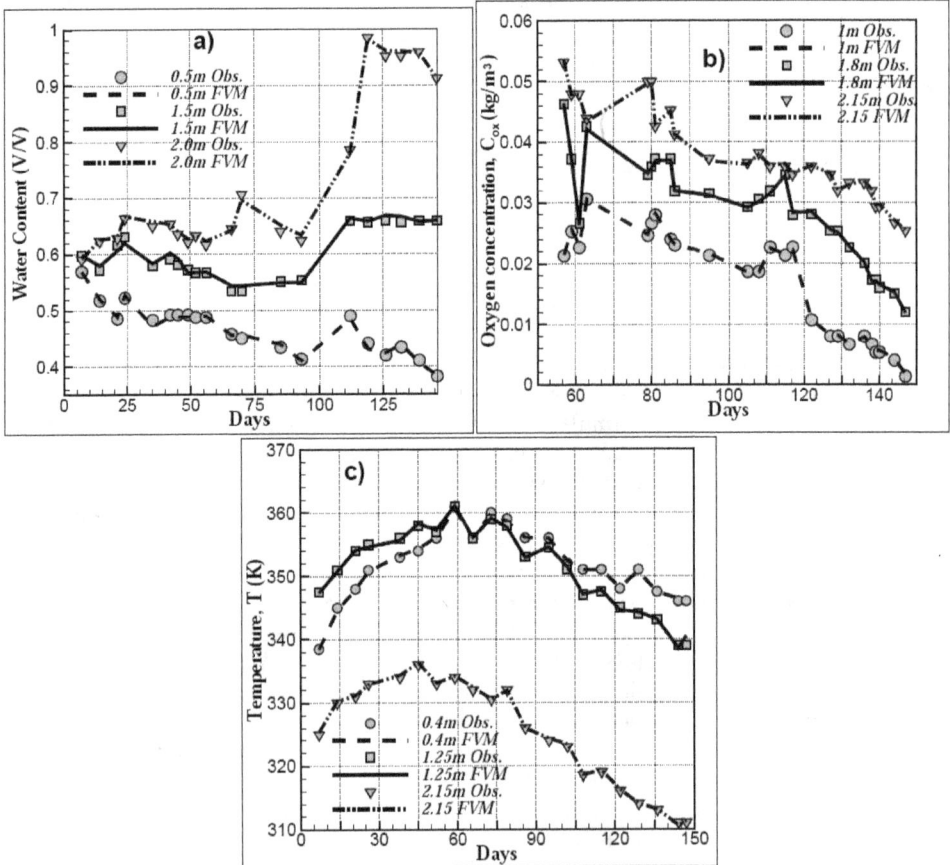

Fig. 14. Water content, oxygen concentration, and temperature observed in the field and calculated with the Finite Volume Methods (FVM) at three heights.

conditions outside the compost pile; at this height the water content is mainly affected by flow into the soil at the pile's base.

In Figure 14.b the oxygen concentration has a tendency to decrease with time. After 115 days the frequency and quantity of rain increase, producing further declines in oxygen concentration because the water displaces the oxygen in the pores. No self ignition conditions were reached in the field during the 21 weeks of the experiment, as shown in Figure 14.c. During the first weeks, temperature in the sewage sludge piles increased up to about 363 K, and it was higher at the first two heights measured within the compost pile. As expected, when temperature in the pile increases, oxygen (C_{ox}) and water content (θ) decrease. Self heating in the compost pile is clearly affected by atmospheric temperature, solar radiation, wind, relative humidity, and precipitation conditions, however further declines in the values are seen after 85 days, caused by the increase in relative humidity and precipitation.

The environment in which the microorganism and chemical reactions occur is altered because of the changes in the moisture and oxygen concentrations, so biological metabolism and chemical reactions decrease, and therefore the temperature within the compost pile also drops. Maximum errors of 0.5K for temperature and 0.0005 m^3/m^3 and kg/m^3 for water content and oxygen concentration between experimental and numerical results were found.

1.13 Conclusions

Numerical simulations indicate that self-combustion does not take place when the piles are less than 1.8 m high, as has been observed in practice. The heat transfer results show that the heating process is initiated by the volumetric heat generation by micro-organisms, and the thermal explosion is caused by cellulose oxidation when the volume to area ratio exceeds 1. The time required to initiate self combustion is inversely proportional to pile height. The internal distribution of temperature and oxygen concentration depends on the geometry of the compost pile. A mathematical model that considers moisture, oxygen and temperature and their corresponding boundary conditions for modeling the compost processes in static compost pile has been proposed.

Numerical simulation with a mesh of 300x300 nodes and dynamic time states between 300 and 3600 s can be used with the Finite Volume Method to predict temperature, oxygen concentration, and humidity within the compost pile.

2. Trace metal leaching in volcanic soils after sewage sludge disposal.

2.1 Introduction

Sewage sludge is the inevitable end product of municipal wastewater treatment processes worldwide. As the wastewater is purified, the impurities removed from the water stream are concentrated. The sludge stream thus contains many chemical and microbiological constituents usually in concentrated forms that may become potential sources of pollutants when the material is released. No matter how many treatment steps it undergoes, at the end, the sludge and/or its derivatives (such as sludge ash) require ultimate disposal, for which the sewage sludge may be land applied, land filled, incinerated, or ocean dumped. There is no entirely satisfactory solution and all of the currently employed disposal options have serious drawbacks. Land application however is by far the most commonly used method around the world. Approximately six million dry tons of sewage sludge are produced annually in the United States (Bastian, 1997). A recent report showed that the annual production of sewage sludge in member countries of the European Union may reach as much as 8×10^6 tons (Bonnin, 2001). Significant amounts of sewage sludge produced in the United States and the western European nations have been applied on land. Depending on the regions, 24 to 89% of the sludge produced in the U.S. (Bastian, 1997)has been applied on land. Bonnin (Bonnin, 2001) reported that 65% of the sewage sludge in France was land applied. The situations in other parts of the world are expected to be similar.

As the residue of municipal wastewater treatment, sewage sludge represents the aggregation of organic matter, pathogens, trace elements, toxic organic chemicals, essential plant nutrients, and dissolved minerals originally dispersed in the wastewater that are captured and transformed by the wastewater treatment processes. Properly managed, the potential pollutants are assimilated via the biochemical cycling processes of the receiving soils in the land application. The practice provides soils with organic materials and offers the possibility of recycling plant nutrients, which in turn improve the fertility (Walter & Cuevas, 1999) and physico-chemical properties of agricultural soils (Illera et al., 2006). If not appropriately controlled, the potential pollutants released through land application may degrade the quality of downstream water bodies, be transferred through the food chain to harm the consumers of harvests, and drastically alter the physical and chemical properties of the receiving soils. It is imperative for mass input to provide adequate amounts of substances that are useful for plant development and for pollutant inputs to be controlled to avert detrimental public health and environmental effects. Major countries such as the United States, the European Union (www.europa.eu.int/comm/environment/sludge) have enacted regulations or issued guidelines that limit the disposal options for a variety of reasons. As already mentioned, municipal sewage sludge contains organic matter, essential plant nutrients, and dissolved minerals, and has buffering capacity (Eriksson, 1998; Zhang et al., 2002a, 2002b; Escudey el al., 2004a, 2004b; Pasquini & Alexander, 2004). When land-applied, they may replenish the depleting nutrient reservoirs in these soils under cultivation, allowing the recovery of soil organic matter lost either during a forest fire or in degradation processes due to adverse environmental conditions and unsuitable agricultural practices (Margherita el al., 2006), but they may also involve the input of variable quantities of heavy metals.

In the sewage sludge used, the levels of heavy metals follow the sequence Zn>Mn>Cu>Cr>Pb>Ni>Mo>Cd (from 1780 mg/kg for Zn down to 5 mg/kg for Cd), with land application ass one of the primary options under consideration at this time. In this sense the evaluation of the total metal content in soils or sewage sludge is useful for a global index of contamination, but it does not provide information about pollutant chemical fractions. On the other hand, it has been widely recognized through biochemical and toxicological studies that the environmental impact of heavy metal pollution can be related to soluble and exchangeable fractions that determine bioavailability, mobility, and toxicity in soils (Rauret, 1998; Lock & Janssen, 2001; Guo et al., 2006a). In soils with a mineralogy dominated by crystalline compounds and with lower organic matter content than volcanic soils, it has been found that a negligible movement of trace metals through the soil profile occurred after 17 years of sludge application (Sukkariyah et al., 2005), and Chang (Chang et al., 1984) found that >90% of metals such as Cd, Cr, Cu, Ni, Pb, and Zn added in that way remained in the surface layer (0-0.15 m) after 6 years. Unlike others contaminants, most metals do not undergo microbial or chemical degradation in the soil; therefore, metal concentrations will remain without significant changes for long periods of time (Guo et al., 2006b).

2.2 Impact on Soils from sewage sludge

In Chile, the treatment works are gradually being brought online in recent years. Before that the collected wastewater was discharged directly and sewage sludge did not exist. With the start of wastewater treatment, sewage sludge and ash from the incinerated sewage sludge are accumulating in the metropolitan areas awaiting final disposal. On the other hand, the soils that would be most affected by these amendments are, of course, those that represent about 70% of the country's arable land, soils derived from volcanic ash. The predominant minerals of these soils are allophane and ferrihydrite in the andisols and kaolinite, halloysite and iron oxides in the ultisols. These soils are rich in iron oxides and organic matter, and they have pH-dependent variable surface charge and high PO_4 accumulation. However, the soils have poor fertility; at the original acidic pH range of 4.5 to 5.5 they have low capacity for exchangeable cations (CEC) and a strong selectivity for K and Ca over Mg (Escudey et al., 2004b). Phosphorus is strongly fixed by the minerals, and therefore it is not readily available for plant absorption in these soils. To be productive, they require frequent adjustments of pH, replenishment of exchangeable Mg, and heavy PO_4 applications. When soil pH increases the CEC increases, P fixation decreases, and K selectivity is reduced. On the other hand, when the soil's organic matter increases, K selectivity is also reduced (Escudey et al., 2004b).

In relation to the impact of biosolids, either in their initial state or as ash, studies in pots and columns have been made on soils derived from volcanic materials. In this sense, forest fires are frequent in central-southern Chile; high temperatures may affect heavy metal (Cu, Zn, Ni, Pb, Cd, Mo, Cr, and Mn) chemical fractions naturally present in the soils and those coming from sewage sludge amendment. Changes in exchangeable, sorbed, organic, carbonate, and residual heavy metal fractions, evaluated by sequential extraction, were observed after heating at 400°C in two amended volcanic soils. The most significant heavy metals in these samples were Cu, Zn, Pb, and Ni. A significant increase in the total content of organic matter and metal ions such as Zn and Cu was observed in amended soils with respect to controls. In all samples, sorbed and exchangeable forms represent less than 10% of the total amount, while organic and carbonate fractions represent 24% and 48%, respectively. The thermal treatment of amended soil samples results in a redistribution of the organic fraction, mainly into more insoluble carbonate and residual fractions such as oxides. Finally, the thermal impact is much more important in soils amended with sewage sludge if a heavy metal remediation process is considered, reducing the mobility and solubility of heavy metals supported by sewage sludge, minimizing leaching, and promoting accumulation in surface horizons (Antilen et al., 2006).

Column leaching experiments were conducted to test the ability of Chilean volcanic soils to retain the mineral constituents and metals in sewage sludge and sludge ash incorporated into the soils. Small or negligible amounts of the total content of Pb, Fe, Cr, Mn, Cd, and Zn (0 to <2%), and more significant amounts of mineral constituents such as Na (7 to 9%), Ca (7 to 13%), PO_4 (4 to 10%), and SO_4 (39 to 46%) in the sludge and sludge ash were readily soluble. When they were incorporated on the surface layer of the soils and leached with 12 pore volumes of water over a 3 month period, less than 0.1% of the total amount of heavy metals and PO_4 in the sludge and sludge ash were collected in the drainage water. Cation exchange selectivity, specific anion adsorption and solubility are the processes that cause the reduction of leaching. The volcanic soils were capable of retaining the mineral constituents, P, and metals in applied sewage sludge and sludge ash and gradually released them as nutrients for plant growth.

2.3 Soil description and methods studied
2.3.1 Soils characterization
Soil samples were collected in southern Chile from a depth of 0 to 25 cmin the areas of Collipulli, Ralún, Diguillín, Metrenco, and Nueva Braunau, reflecting the localities from which the soils were extracted. The samples were obtained from well drained and regularly cultivated fields. Collipulli and Metrenco are classified as ultisols and Ralún, Diguillin, and Nueva Braunau as andisols. General information on the climate and geography of the soils is given in Table 4. Also, mineralogical composition can be observed in Table 5. The soil samples were screened in the field to pass a screen with 2 mm openings and stored at the field moisture content in a 4°C cold room until used.

Soil	Location		Soil Order	Classification	Altitude	Rainfall	Annual Mean Temperature
	Longitude	Latitude			(m)	(m yr^{-1})	(°C)
Collipulli (C)	36°58'S	72°09'W	Ultisol	Fine, mesic, Xeric, Paleumult	120 - 400	1.2 - 1.5	15.8
Metrenco (M)	38°34'S	72°22'W	Ultisol	Fine, mesic, Paleumult	100 - 300	1.2 - 1.5	14.6
Ralun (R)	41°32'S	73°05'W	Andisol	Mesic, Umbric Vitrandept	600 - 1,400	4.0 - 5.0	10.5
Diguillin (D)	36°53'S	72°10'W	Andisol	Medial, thermic, Typic Dystrandept	120 - 180	1.2 - 1.8	15.5
Nueva Braunau (NB)	41°19'S	73°06'W	Andisol	Ashy, mesic Hydric Dystrandept	100 - 150	1.2 - 1.5	11.5

Table 4. Soil classification information

Mineral					
	C	M	R	D	NB
Allophane			+++++	+++++	+++++
α-Cristobalite	+		+		+
Chlorite - Al			++		
Feldspars					+
Ferrihydrite				+	+
Gibbsite					++
Geothite		+			
Halloysite	+	+++++		++	
Kaolinite	+++++				
Montmorillonite			+		
Organo-allophanic				+	++
Plagioclase			++	++	
Quartz		+			
Vermiculite	+			+	

+++++ represents dominant (>50%),
++++ represents abundant (20-50%),
+++ represents common (5 - 20%),
++ represents present (1 - 5%), and
+ represents trace fraction (<1%)

Table 5. Mineralogical composition of soils as represented by the B horizon.

2.3.2 Column experiments
Soils were packed to a depth of 25 cm in 30 cm long and 10 cm diameter acrylic columns, according to their respective field bulk densities. A filter paper disk was placed on the perforated plate at the bottom of each column to prevent the loss of solid materials. The sewage sludge was obtained from a domestic water treatment plant located in Santiago (Chile) and the sewage sludge ash was obtained by heating the sewage sludge at 500°C for

two hours. Depending on the treatment, 30 g of air-dried sewage sludge or the ash equivalent of 30 grams of air-dried sewage sludge were placed on the surface 5 cm of the packed columns. The experimental controls received no sludge or ash treatment. The columns, placed vertically, were flooded once a week with one pore volume of distilled water, and drained by gravity from top to bottom, for a period of 12 weeks. Furthermore, 30 g of sludge and the ash equivalent of 30 g of sludge were leached in the same manner. The drained liquid from each weekly leaching cycle was analyzed for pH, electric conductivity, SO_4, PO_4, Na, K, Mg, Ca, Zn, Cu, Fe, Al, Ni, Cd, Pb, Mo, and Mn.

At the end of the leaching experiment, each soil column was cut open lengthwise and the profile was sectioned into five equal length segments for analysis of the soils' pH, electric conductivity, and organic carbon, exchangeable cation and P contents. A chemical fractionation of heavy metals was carried out in sludge and sludge ash using the methodology proposed by Chang (Chang et al., 1984). Sequential extraction with 0.5M KNO_3, distilled water, 0.5M NaOH, 0.05M EDTA, and 0.5M HNO_3 allowed the estimation of the exchangeable, sorbed, organic, carbonate, and residual heavy metal fractions.

2.3.3 Chemical determinations

The bulk density, exchangeable cations, total porosity, and organic carbon content of the soils were determined by methods outlined in Methods of Soil Analysis. Briefly, bulk density (Blake, 1965) was determined by the average air-dried weight of soils in undisturbed soil cores of the 0 to 25 cm soil profile in 5 cm (diameter) x 5 cm (height) brass rings; exchangeable cations were determined as the concentrations of Na, K, Mg, and Ca in ammonium acetate extracts (Peech, 1965); and organic carbon was determined by the Walkley-Black method (Allison, 1965). The pH and electric conductivity of the soils were measured in soil suspensions with a1: 2.5 w/v soil-to-water ratio. The total elemental contents of Na, K, Mg, Ca, Zn, Cu, Fe, Al, Ni, Cd, Pb, Mo, Mn, P and S were determined by digesting the soils with a concentrated HNO_3-HCl-HF mixture in a microwave oven and measuring the concentrations by ICP-OES spectroscopy (Perkin Elmer Optima 2000 equipment). Comparable components of the sewage sludge and sludge ash were determined in the same manner. The concentration of the same elements in leachates was also determined by ICP-OES; SO_4 and PO_4 concentrations in the drainage water were measured by ion chromatography (Waters 625LC) in a Waters IC Pak anion HR 4.5x75 mm column. The absorbance of leachates was measured at 465 and 665 nm in a UV-Visible Perkin Elmer Lambda 20 spectrophotometer.

Prior to the sludge and ash treatments, the soils were acidic, with pH varying between 4.5 and 5.9, and low in exchangeable base contents varying from 1.5 to 10.4 cmol kg^{-1} (Table 6). In contrast, the sewage sludge and sludge ash had pH 7.7 and 7.4, respectively, 2 to 3 orders magnitude higher in alkalinity than the soils. The exchangeable base content of the sewage sludge was 80.6 cmol kg^{-1}, 10 to 54 times higher than that of the soils. The Na, K, Mg and Ca in the sludge ash were soluble but not necessarily in the exchangeable forms. Judging from their electric conductivities, the soluble mineral contents of sewage sludge and sludge ash were orders of magnitude higher than those of the soils, even though the incineration of sewage sludge results in less soluble chemical forms, and consequently presents a lower electric conductivity than the sewage sludge. The total elemental Ca, Mg, K, and Na content in the soils follows the same trends as that in the exchangeable forms and the concentrations are in the same order of magnitude. Column pore volume was calculated considering the amount of soil in the column and the total porosity of each soil (Table 6).

Soil	pH	Bulk Density (g cm⁻³)	Pore Volume (mL)	Organic Carbon (%)	Electrical Conductivity (µS m⁻¹)	Exchangeable Bases (cmol kg-1)			
						Na	K	Mg	Ca
Collipulli (C)	5.4	1.36	1027	2.3	81	0.1	0.2	1.8	5.9
Metrenco (M)	5.5	1.33	1056	1.8	29	0.2	0.3	1.5	4.0
Ralún (R)	4.5	0.90	988	6.2	436	0.1	0.1	0.4	2.5
Diguillin (D)	5.9	1.12	830	6.5	94	0.2	0.7	1.1	8.4
N.Braunau (NB)	5.5	0.82	834	11.0	20	0.1	0.1	0.2	1.1
Sludge	7.7	0.46	-	17.8	8520	1.5	2.5	10.7	65.9
Sludge Ash	7.4	-	-	<0.1	3890	1.2	1.1	7.4	25.8

Table 6. Properties of soils, sewage sludge and sludge ash.

2.1.4 Releases from sludge and sludge ash

When the sludge and sludge ash were leached, soluble species such as K, Na, Ca and Mg appeared in the leachates. In general, the behavior observed for the K, Na and Mg species indicates a gradual and constant elution, with an important removal in the first pore volume, considering that the curves describe the accumulated amount of exchangeable bases. Comparatively, Fig. 15 shows greater elution from the sludge than from the sludge

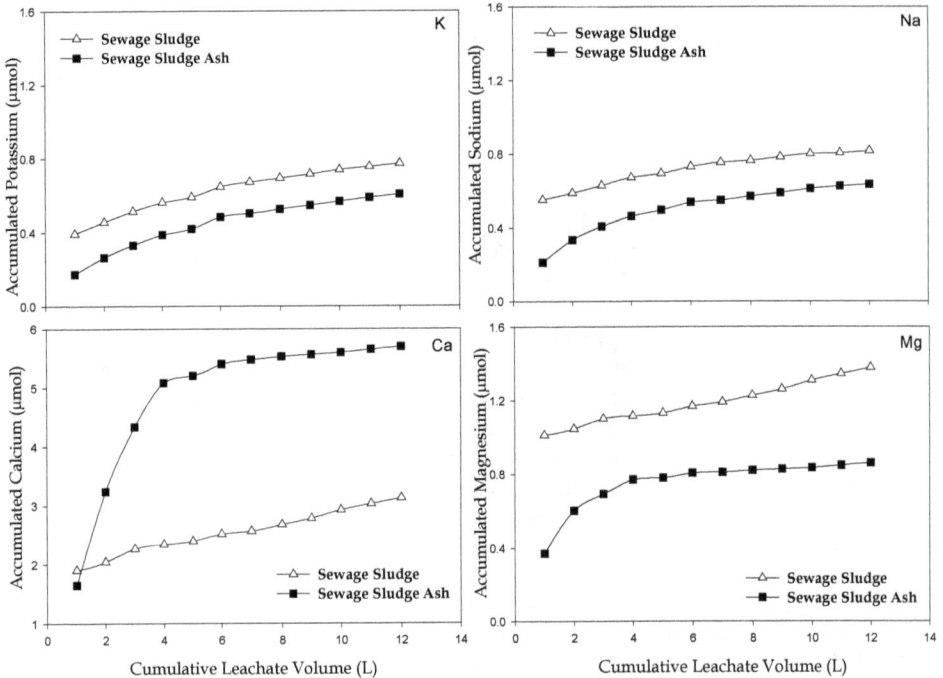

Fig. 15. Accumulated exchangeable bases (K, Na, Ca and Mg) from sewage sludge and sludge ash.

ash, except for Ca. In that relation Ca also presents the greatest elution in the first four pore volumes, exceeding largely the elution from the sludge. This behavior is related to the addition of lime that is made in the water treatment plants with the purpose of stabilizing the pH of the residues. Another species of interest is sulfate, where the soluble SO_4 in sewage sludge was depleted with one pore volume of water used to leach the soils. In contrast, the soluble SO_4 in sewage sludge ash is gradually released with 5 to 8 pore volumes of water, with total amounts released of 342 and 319 mg, respectively.

One main domain is observed in sludge release which is associated to highly soluble forms. On the other hand, two main domains are observed in sewage ash, the first associated with soluble forms which are less important than in sludge, and a second from 2 to 5 pore volumes which can be associated with slow equilibrium between solid and water. In both samples the quantities released were a small fraction of the total amounts.

Only small amounts of K, Na, Ca, Mg and SO_4 were released when the sludge and sludge ash were subject to intense leaching for 12 weeks.

In Chile, the total metal content in the sewage sludge follows the sequence Zn > Cu > Pb > Ni. Fractionation data show that Zn and Cu are mainly associated with highly insoluble fractions, such as carbonates and residual fraction. In control soils the total heavy metal content follows the sequence Zn > Cu > Ni > Pb for Collipulli soil, and Cu > Zn > Ni> Pb for Ralún soil (Antilen et al. 2006). On the other hand, the Zn and Cu release patterns for the sludge and sludge ash were similar (Figure 16), with the accumulated amounts released by the sludge considerably higher than those of the sludge ash.

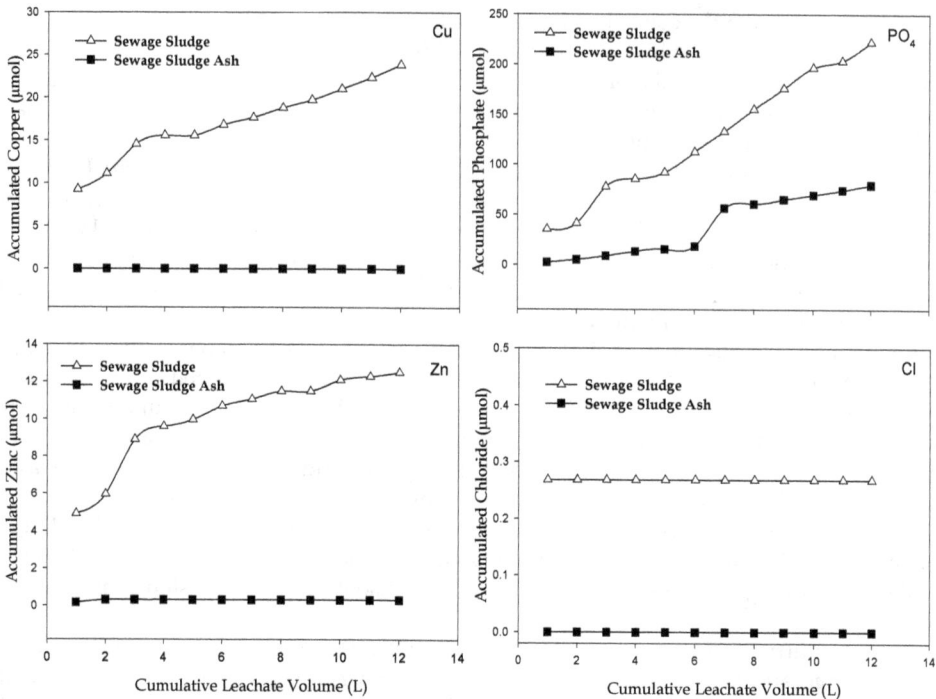

Fig. 16. Accumulated releases of heavy metals (Cu, Zn), phosphorus (PO_4) and chloride (Cl).

In relation to organic and inorganic P forms, both are present in sludge, while in sludge ash only inorganic P forms are present. The P forms in both samples are released slowly and at constant rates over time. In sludge, release is probably controlled by slow equilibrium between solid organic P forms and soil solution, and by the solubility of inorganic P forms. Consequently, at the end of 12 leaching cycles, small amounts of PO_4 were recovered from the drainage liquids of sewage sludge and sludge ash (18 and 6 mg, respectively) compared with their total contents (181 and 170 mg, respectively).

Even though Cu and Zn are the main heavy metals in Chilean sewage sludge, other heavy metals of environmental interest, such as Ni, Cd, Cr, Mo and Mn, were also considered. The total eluted amounts of some of these metals are shown in Table 7, where it is clear that it is minimal compared to the content in the sludge.

	Total leached amount (μmol)						
	Cr	Ni	Mo	Mn	Cd	Cu	Zn
Sewage sludge	1.00	6.07	1.61	9.15	0.00	0.00	0.789
Sewage sludge ash	0.32	0.11	2.47	1.86	0.00	1.61	0.02

Table 7. Accumulated heavy metals leached with12 pore volumes from sewage sludge and sludge ash

2.4.1 Soil attenuation

The pH of leachates in the control and treated soils increases after 12 pore volumes; the final pH is about 1.5 to 2.0 units higher than the initial pH. The process is controlled by the soil; thus, after 12 pore volumes the pH of treated soil leachates is only about 0.3 pH units higher than those observed in the control columns. In all the experiments, after 12 pore volumes, the leachate pH is basic, ranging from 7 to 8.

The leaching of organic matter was followed by measuring the absorbance of the leachates after each pore volume at 465 and 665 nm. Only leachates from Ralún soil columns showed absorbance higher than zero, but the amount of organic matter leached was too low to be quantified. No significant loss of organic colloids was observed, because the mass balance shows that the organic carbon remains constant in all columns considering the experimental errors of the Walkley-Black method.

Even without the applications of sludge or sludge ash, cations and anions such as Mg, and SO_4 may be leached from the soils (Figure 17) and the amounts collected in the drainage water were dependent on the conditions of the soils. Sludge and sludge ash amendments consistently enhanced the leaching of minerals. However, the collected amounts were significantly smaller than the total introduced through the addition of sludge or sludge ash, and are practically leached in the first 3 or 4 pore volumes of drainage water.

Soil incorporation further reduced the mobility of the chemical constituents in the sludge and sludge ash (Figure 18). For P, the amounts found in the drainage water (Figure 18) were 2 to 3 orders of magnitude lower than the amounts present in the added sludge and sludge ash.

As a result, nutrients such as available P significantly increased with the application of sewage sludge and sludge ash for both the Ultisol and Andisol (Figure 19). The general trend in all the experiments was that only a small fraction of the total amounts incorporated by the addition of sludge or sludge ash were leached.

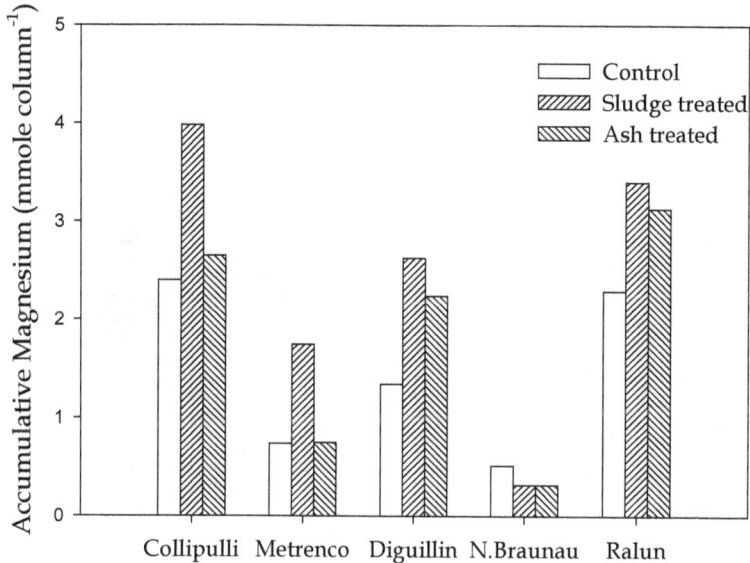

(A)

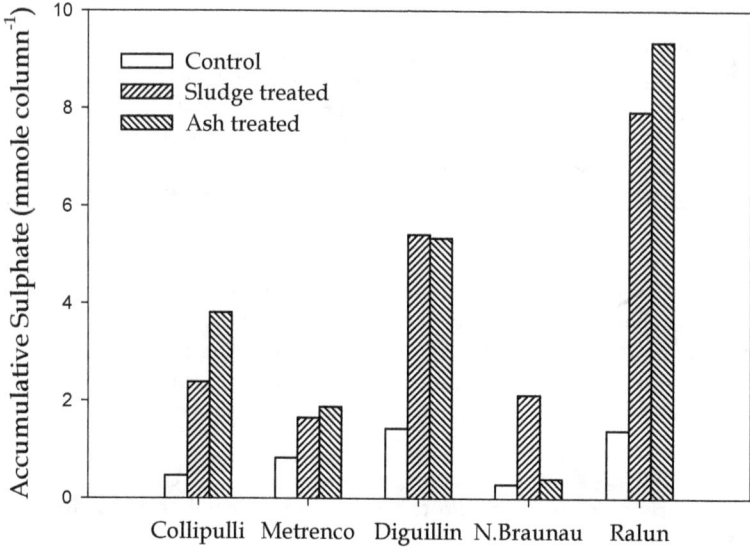

(B)

Fig. 17. Total amount of Mg (A) and SO_4 (B) leached from sewage sludge and sludge ash treated soils.

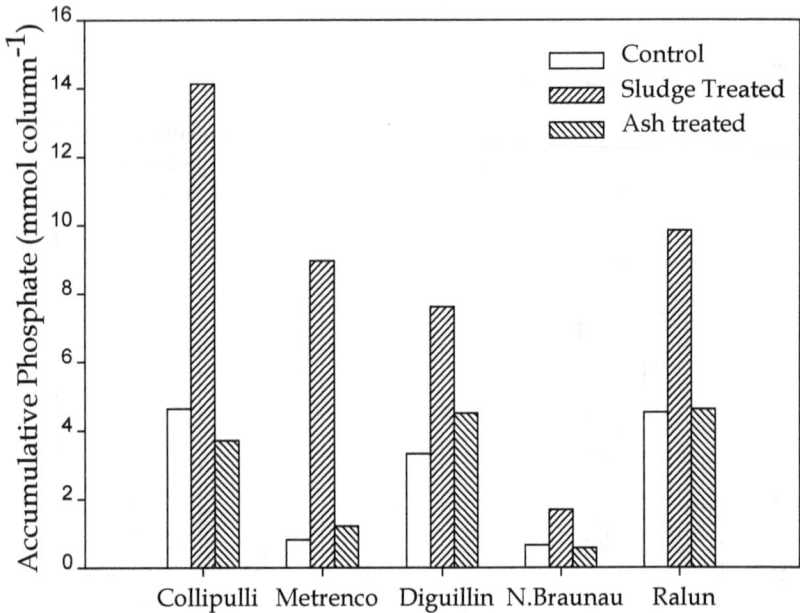

Fig. 18. Total amount of PO$_4$ leached from sewage sludge and sludge ash treated soils.

As an example, the total input from sludge and ash, the total amount leached from them, and the total amount collected after 12 pore volumes for the Collipulli and Nueva Braunau soils, are presented in Figure 20. The total amount of heavy metals (Cu, Zn, Ni, Cd, Pb, Mo, Mn) leached after 12 pore volumes was <0.1% of the total input from sewage sludge or sewage ash (represented by Zn, Cu and Pb in Figure 20). On the other hand, the leached fractions of SO$_4$ (22 to 55%), Na (7 to 15%) , K (2 to 30%), Ca (3 to 7%), and Mg (2 to 30%) are more significant.

The leaching of exchangeable bases behaves as predicted by previously reported cation exchange selectivity (Escudey et al. 2002). Phosphate is leached in very low amounts (<0.1%), even though sewage sludge and sludge ash present high P content; this is due to the specific PO$_4$ adsorption which is a characteristic of Chilean volcanic soils (Escudey et al. 2001).

Fractionation experiments show that 86 to 99% of heavy metal chemical forms in sewage sludge are associated with organic matter complexes, carbonate, and residual low solubility compounds, and that 95 to 99% is associated with carbonate and low solubility forms in sludge ash. All of them have low mobility, and consequently their leaching is mainly associated to the more soluble chemical forms, which are present only in very low concentration in both substrates.

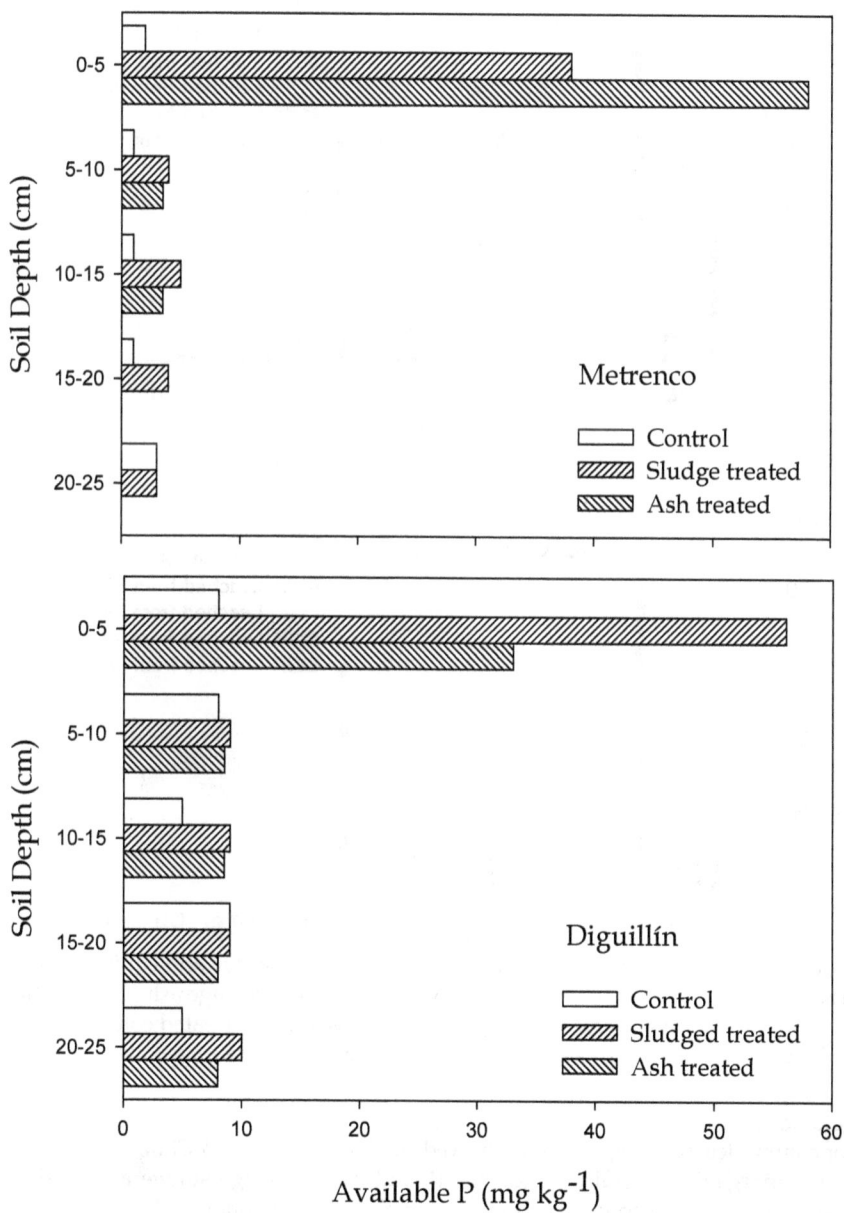

Fig. 19. Available P in the sewage sludge and sludge ash treated ultisol (Metrenco) and andisol (Diguillin).

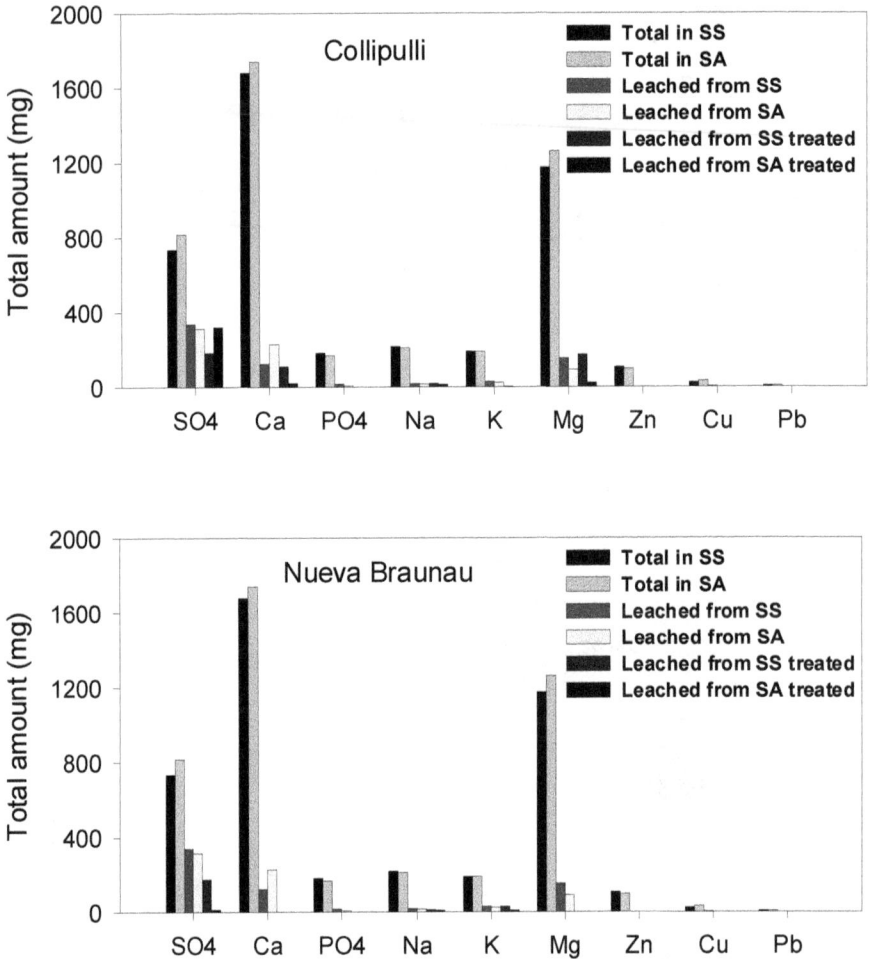

Fig. 20. Total amount of selected cations and anions in sewage sludge and equivalent ash (Total in SS, SA), total amount leached from sewage sludge and sewage ash (Leached from SS, SA), and leached from sewage sludge-treated columns and ash-treated columns (Leached from SS treated, ash treated), for Collipulli and Nueva Braunau soils.

2.5 Conclusions

Results of column leaching experiments showed that volcanic soils in Chile were capable of retaining the inorganic mineral constituents, P, and Zn in sewage sludge and sludge ash when land applied. These constituents are essential inputs to enhance the productivity of volcanic soils that are frequently low in fertility. Cation exchange selectivity, specific anion adsorption and solubility are the processes that cause the reduction of leaching. In this regard, the volcanic soils will attenuate the sewage sludge-borne pollutants and provide soils with nutrients that may be slowly released for crop production.

3. References

Allison, L. E. (1965). Organic carbon In: *Methods of Soil Analysis. Part 2* Agronomy 9, Black, C.A.; Evans, D.D.; White, J. L.; Ensminger, L.E. & Clark, F.E. (Eds.),1367-1396, American Society of Agronomy, Madison, WI.

Antilen, M.; Araya, N.; Briceño,M.; Escudey, M (2006) Changes on chemical fractions of heavy metals in Chilean soils amended with sewage sludge affected by a thermal impact. *Australian Journal of Soil Research* 44, 1-7, 0004-9573.

Bastian, R. K. (1997). Biosolids management in the United States, A state-of-the-nation review. *Water Environmental Technology*, 9, 45 – 50, 1044-9493.

Blake, G.R. (1965). Bulk density, In: *Methods of Soil Analysis. Part 1* Agronomy 9, Black, C.A.; Evans, D.D.; White, J. L.; Ensminger, L.E. & Clark, F.E. (Eds.), 1367-1396, American Society of Agronomy, ISBN, Madison, WI.

Bonnin, C. (2001). Organic pollutants and sludge – The French experience. Research The Sludge Directive, *Proceedings of a Conference on Sewage Sludge,* Belgian, October, 2001, Brussels.

Chang, A.C.;. Warneke, J.E.; Page, A.L.& Lund L. J. (1984). Accumulation of heavy metals in sewage sludge-treated soils. Journal of Environmental Quality, 13, 87-91, 0047-2425.

Chen, X.D. & Mitchell, D.A. (1996). Star-up strategize of self-heating and efficient growth in stirred bioreactor for solid state bioreactors, Proceedings of the 24th Annual Australian and New Zealand Chemical Engineering Conference (CHEMECA 96), 111-116.

Duffie, J.A. & Beckman, W.A. (1980). Solar Engineering of Thermal Processes, John Wiley & Sons, New York, 1980.

Eriksson, J.(1998). Dissolution of hardened wood ashes in forest soils: studies in a column experiment, Scandinavian. *Journal of Forest Research* , 2, 23-32, 1341-6979.

Escudey M, Galindo G, Avendaño K, Borchardt D, Chang A, Briceño M (2004*a*) Distribution of phosphorus forms in Chilean soils and sewage sludge by chemical fractionation and 31P-NMR. *Journal of Chilean Chemical Society* 49, 219-222, 0717-9707.

Escudey M, Forster JE, Galindo G (2004b) Relevance of organic matter in some chemical and physical characteristics of volcanic ashderived soils in Chile. *Communications in Soil Science and Plant Analysis* 35, 781-797, 1532-2416.

Escudey M.; Arias, A.; Förster, J.; Moraga, N.; Zambra, C. & Chang, A. C. (2008). Sewage sludge self-heating and spontaneous combustion. field, Laboratory and numerical studies; High Temperature Material and Processes 27(5).

Escudey, M.; Diaz, P.; Förster, J.E.; Pizarro, C.; Beltrán, L. & Galindo, G.(2002). Prediction of K-Ca-Mg ternary exchange from binary isotherms in volcanic soils using the Rothmund-Kornfeld approach. Australian Journal of Soil Research, 40, 781-790,0004-9573.

Escudey, M.; Galindo, G.;. Förster, J. E.; Briceño, M.; Díaz, P.; Chang, A. C. (2001). Chemical forms of phosphorus of volcanic ash-derived soils in Chile, Communication in Soil Science and Plant Analysis, 32, 601-616, 1532-2416.

Geankoplis, C.J. (1993). Transport processes and unit operations, Prentice–Hall International Inc. USA.

Guo G, Zhou Q, Koval P, Belogolova G (2006a) Speciation distribution of Cd, Pb, Cu, and Zn in contaminated Phaeozem in north-east China using single and sequential extraction procedures. *Australian Journal of Soil Research* 44, 135-142, 0004-9573.

Guo G, Zhou Q. Lena M (2006b) Availability and assessment of fixing additives for the in situ remediation of heavy metal contam United soils: a review. *Environmental Monitoring and Assessment* 116, 513-528, 0167-6369.

http://europa.eu.int/comm/environment/sludge.

Illera, V.; Walter, I.; Souza, P. & Cala, V. (2000). Short-term effects of biosolid and municipal solid waste applications on heavy metals distribution in a degraded soil under a semi-arid environment. *Science of the Total Environment*, 255, 29-44, 0048-9697.

Kaya, A.; Aydin, O. & Dincer, I. (2006). Numerical modeling of heat and mass transfer during forced convection drying of rectangular moist objects, International Journal of Heat andMass Transfer 49, 3094−3103.

Lock K, Janssen C (2001) Cadmium toxicity for terrestrial invertebrates: taking soil parameters affecting bioavailability into account. *Ecotoxicology* 10, 315-322, 0963-9292.

Margherita E, Brunetti G, Garcia-Izquierdo C, Covalcante F, Fiore S, Senesi N (2006) Humic substances and clay minerals in organically-amended semiarid soils. *Soil Science* 171, 322-333, 0038-075X.

Moraga, N.; Corvalán, F.; Escudey, M.; Arias, A. & Zambra, C. (2009). Unsteady 2D coupled heat and mass transfer diffusion in porous media with biological and chemical heat generations; International Journal of Heat and Mass Transfer 52; 5841–5848.

Nield, D. & Bejan, A. (1992). Convection in Porous Media, Springer-Verlag, New York.

Ozisik, M.N. (1994). Finite difference methods in heat transfer, CRC Press Inc., USA.

Pasquini, M.W. & Alexander, M.J.(2004). Chemical properties of urban waste ash produced by open burinig on the Jos Plateau: implications for agriculture. *Science of the Total Environment*, 319, 225–240, 0048-9697.

Patankar, S.V. (1991). Computation of Conduction and Duct flow Heat Transfer, Innovative Research, Inc., Maple Grove, Minnesota.

Peech, M. (1965). Exchange acidity In: *Methods of Soil Analysis. Part 1* Agronomy 9, Black, C.A.; Evans, D.D.; White, J. L.; Ensminger, L.E. & Clark, F.E. (Eds.),   905-913, American Society of Agronomy, Madison, WI..

Rauret G (1998) Extraction procedures for the determination of heavy metals in contaminated soils and sediment. *Talanta* 46, 449-455, 0039-9140.

Richards, L.A. (1931). Capillary conduction of liquids through porous mediums, Physics 1, 318−333.

Serrano, S. (2004). Modeling infiltration with approximated solution to Richards equation, Journal of Hydrologic Engineering 9(5), 421−432.

Sidhu, H.S.; Nelson, M.I. & Chen, X.D. (2007). A simple spatial model for self-heating compost piles, ANZIAM J. (CTAC2006) 41, C135-C150.

Sukkariyah B, Evanylo G, Zelazny L, Chaney R (2005) Recovery and distribution of biosolids-derived trace metals in a clay loam soil. *Journal of Environmental Quality* 34, 1843-1850, 0047-2425.

Walkley, A. & Black, L.A. (1934). An examination of the Degtjareff method for determining soil organic matter and a proposed modification of the chromic acid titration method. Soil Science 37, 29–38.

Walter, I. & Cuevas, G. (1999). Chemical fractionation of heavy metals in a soil amended with repeated sewage sludge application. *Science of the Total Environment*, 226, 113 – 119, 0048-9697.

Zhang, F.; Yamasaki, S.; Nanzyo, M. (2002*a*). Waste ashes for use in agricultural production: I. Liming effect, contents of plant nutrients and chemical characteristics of some metals. *Science of the Total Environment*, 284, 215–225, 0048-9697.

Zhang, F.; Yamasaki, S.; Nanzyo, M. (2002b). Waste ashes for use in agricultural production: II. Contents of minor and trace metals. *Science of the Total Environment*, 286, 111–118, 0048-9697.

Permissions

All chapters in this book were first published in WWEM, by InTech Open; hereby published with permission under the Creative Commons Attribution License or equivalent. Every chapter published in this book has been scrutinized by our experts. Their significance has been extensively debated. The topics covered herein carry significant findings which will fuel the growth of the discipline. They may even be implemented as practical applications or may be referred to as a beginning point for another development.

The contributors of this book come from diverse backgrounds, making this book a truly international effort. This book will bring forth new frontiers with its revolutionizing research information and detailed analysis of the nascent developments around the world.

We would like to thank all the contributing authors for lending their expertise to make the book truly unique. They have played a crucial role in the development of this book. Without their invaluable contributions this book wouldn't have been possible. They have made vital efforts to compile up to date information on the varied aspects of this subject to make this book a valuable addition to the collection of many professionals and students.

This book was conceptualized with the vision of imparting up-to-date information and advanced data in this field. To ensure the same, a matchless editorial board was set up. Every individual on the board went through rigorous rounds of assessment to prove their worth. After which they invested a large part of their time researching and compiling the most relevant data for our readers.

The editorial board has been involved in producing this book since its inception. They have spent rigorous hours researching and exploring the diverse topics which have resulted in the successful publishing of this book. They have passed on their knowledge of decades through this book. To expedite this challenging task, the publisher supported the team at every step. A small team of assistant editors was also appointed to further simplify the editing procedure and attain best results for the readers.

Apart from the editorial board, the designing team has also invested a significant amount of their time in understanding the subject and creating the most relevant covers. They scrutinized every image to scout for the most suitable representation of the subject and create an appropriate cover for the book.

The publishing team has been an ardent support to the editorial, designing and production team. Their endless efforts to recruit the best for this project, has resulted in the accomplishment of this book. They are a veteran in the field of academics and their pool of knowledge is as vast as their experience in printing. Their expertise and guidance has proved useful at every step. Their uncompromising quality standards have made this book an exceptional effort. Their encouragement from time to time has been an inspiration for everyone.

The publisher and the editorial board hope that this book will prove to be a valuable piece of knowledge for researchers, students, practitioners and scholars across the globe.

List of Contributors

Eduardo Robson Duarte, Flávia Oliveira Abrão, Neide Judith Faria de Oliveira and Bruna Lima Cabral
Instituto de Ciências Agrárias da Universidade Federal de Minas Gerais Montes Claros, Brazil

Masoud Tabari, and Azadeh Salehi
Tarbiat Modares University, Iran

Jahangard Mohammadi
Shahrekord University, Iran

Alireza Aliarab
Gorgan University, Iran

Dr. Kelly T. Morgan
University of Florida, Soil and Water Science Department, Southwest Florida Research and Education Center, Immokalee, FL 34142, USA

Akuzuo Ofoefule, Eunice Uzodinma and Cynthia Ibeto
Biomass Unit, National Center for Energy Research & Development, University of Nigeria, Nsukka Enugu state, Nigeria

Owen Fenton, Mark G. Healy, Raymond B. Brennan, Ana Joao Serrenho, Stan T.J. Lalor, Daire O hUallacháin and Karl G. Richards
Teagasc, Environmental Research Centre, Wexford, National University of Ireland, Galway, Rep. of Ireland

Alex C. Chindah and Solomon A. Braide
Institute of Pollution Studies, Rivers State University of Science and Technology, PMB 5080, Port Harcourt, Nigeria

Charles C. Obunwo
Chemistry Department, Rivers State University of Science and Technology, PMB 5080, Port Harcourt, Nigeria

Peace Amoatey (Mrs) and Professor Richard Bani
Department of Agricultural Engineering, Faculty of Engineering Sciences, University of Ghana, Ghana

G.M. Cappelletti, G.M. Nicoletti and C. Russo
Dipartimento SEAGMeG – University of Foggia, Italy

F. Sanem Sunlu, Ugur Sunlu, Baha Buyukisik, Serkan Kukrer and Mehmet Aksu
Ege University, Faculty of Fisheries, Dept. of Hydrobiology, Turkey

Nelson Moraga and Carlos Zambra
Universidad de Santiago de Chile, Facultad de Ingeniería, Departamento de Ingeniería Mecánica, Av. Bdo. O'Higgins 3330, Chile

Mónica Antilén
Pontificia Universidad Católica de Chile, Facultad de Química, Departamento de Química Inorgánica,Vicuña Mackenna 4860, Santiago, Chile
Center for the Development of Nanoscience and Nanotechnology, CEDENNA, Santiago, Chile

Mauricio Escudey
Universidad de Santiago de Chile, Facultad de Química y Biología, Departamento de Química de los Materiales, Av. B. O'Higgins 3363, Santiago, Chile
Center for the Development of Nanoscience and Nanotechnology, CEDENNA, Santiago, Chile

Index

www.ingramcontent.com/pod-product-compliance
Lightning Source LLC
Chambersburg PA
CBHW050451190326
41458CB00005B/1231